Digital Social Innovation

Chiara Certomà

Digital Social Innovation

Spatial Imaginaries and Technological Resistances in Urban Governance

palgrave
macmillan

Chiara Certomà
Department of Economic, Social
Mathematical and Statistical Sciences
University of Turin
Turin, Italy

ISBN 978-3-030-80450-3 ISBN 978-3-030-80451-0 (eBook)
https://doi.org/10.1007/978-3-030-80451-0

This Palgrave Macmillan imprint is published by the registered company Springer Nature Switzerland AG
The registered company address is: Gewerbestrasse 11, 6330 Cham, Switzerland

Acknowledgements

This book results from several pleasant and unexpected encounters that allowed me to find inspiration where I least expected it. Particularly, I would like to thank some of the people who helped me explore the digital society in different ways.

The marvellously imperfect city of Ghent and its trouble-making citizens have been a first and permanent source of surprise; and pushed me to look at both the bright and dark sides of the phenomena you will read about in the following pages. Colleagues and friends at the University of Ghent (notably Thomas Block, Michiel Dehaene, Bruno Notteboom and the whole team at the Centre for Sustainable Development—CDO) have been extremely supportive during my EU Marie Skłodowska-Curie Action on "Crowdsourcing Urban Sustainability Governance".

From the other side of the world, Mark Dyer offered me the opportunity to visit The Waikato University to feed up our long-lasting and productive dialogue on contemporary urban narratives.

In Italy I had the opportunity to think through the collected materials with the unconditioned assistance of Filippo Corsini (who also shared with me some of the graphic elaborations included in Chapter 2) and the support of Marco Frey at the Sant'Anna School of Advanced Studies in Pisa. A bit later, my colleagues at the University of Turin and the ESOMAS Department welcomed me in a stimulating intellectual environment. There I met Paolo Giaccaria who enthusiastically engaged with me in analysing the multiple and contradictory faces of social innovation.

Moreover, I want to thank Tanya Rahman, who proofread the book and offered me valuable suggestions and Veronica Vitale, who realised the visual stories reproduced in Chapters 4 and 6. The editorial team at Palgrave Macmillan, Rachael Ballard and Shreenidhi Natarajan, generously assisted me in the final production stage.

Special thanks to my dad, Giuseppe, for patiently helping me with the editing process. This book is dedicated to him, to my mother, Caterina (who never stops looking, in the present, for the seeds of the future) and to Pipo (who wisely chose to do everything the opposite of what I did).

Contents

List of Figures

List of Tables

CHAPTER 1

Introduction

Abstract The chapter explains how I come across the Digital Social Innovation (DSI) issue and why I decided to explore it. Motivated by the curiosity to understand the nature of digital innovation and its social, cultural and political consequences in public life, I share here my reasons for adopting a critical geography approach in the analysis of digital revolution's impact on the city. I recognise that the concept of digital social innovation is characterised by an intrinsic ambivalence that can be regarded as the source of the reluctancy in adopting it among critical scholars. At present, existing research principally falls within the management and regional economics domain. These, however, devoted a limited attention to the socio-political and cultural implications of digital innovation. In the second part of the book, we will consider how relevant the costs associated with such a lack of attention are. In the meanwhile, this chapter explains why we need to adopt a more critical approach towards DSI, able to balance the enthusiasm for potential improvement of life quality brought about by these innovations with attentive consideration of their "dark sides".

Keyword Digital social · Innovation · Social innovation · City · European Commission · Critical geography · Innovation management

C. Certomà, *Digital Social Innovation*,
https://doi.org/10.1007/978-3-030-80451-0_1

1.1 Following Inspiring Practices

Ghent, May 2019: The winning projects of the *BurgerBudget* participatory budget initiative are announced by the evaluation board.

Promoted by the City Council, the initiative saw over 100 submitted proposals from informal groups of citizens who wanted to make their city, or part of it, a better place to live. Submissions were made to a dedicated online platform. To facilitate this, the city library created new free internet access points and a mobile office went through the city for months, helping non-digitally literate citizens post their proposals online. Citizens e-voted upon submitted projects, including, for instance: the creation of mutual help digital platforms; the transformation of derelict squares or buildings into meeting places and the organisation of communal vegetable gardens and markets. A jury of experts evaluated the proposals with the most votes, merging similar proposals to fund as many as projects as possible.

Tallin, March 2020: Estonian start-ups meet online for the hackathon against the economic crisis induced by the Covid19 pandemic.

With the (quasi)totality of social services digitalised and thousands of free hotspots (including in the forest), Estonia is one of the most wired countries in Europe. Recalling the *Eurocity* association's mantra "If we get it right in cities, we will get it right for Europe", the Tallin hackathon aimed to address the consequences of the pandemic-induced economic crisis. Estonian president, Kersti Kaljulaid, invited everybody to "Hack the Crisis". More than a hundred projects have been elaborated during the hackathon, from the creation of a voluntary medical personnel database (called *Vaab Mtu*) to a platform re-distributing workforces from businesses not operating to those where workers were needed (called *Share Force One*).

Lisbon, Nov 2020: The free languages platform SPEAK reaches 7800 participants in 24 worldwide cities learning and teaching 19 different languages.

In eight years, the initiative promoted by the Portuguese non-profit organisation *Associação Fazer Avançar* in Lisbon to help migrant people, foreigners and refugees to get a fast, hands-on knowledge of local language, has been rapidly growing. Through talk-groups and both online and offline activities, SPEAK facilitates the complex task of increasing social cohesion through the support of a broad network of civic or student associations, hostels and private cultural foundations.

Despite being disparate in place and different in aims, these initiatives have something in common: they all engage people's creativity and collaborative energy to develop digitally based or digitally enabled solutions to collective problems with the primary intent of tackling (and potentially solving) social issues. Most of the proposed solutions root in the urban contexts and embroider on their peculiar cultural and material texture to advance novel perspectives, approaches or just novel tools that (hopefully) improve the organisation of society and the quality of the urban space. Their multifaced, ephemeral character and polymorphous manifestations make it difficult to grasp the intimate nature of these initiatives or to identify their key traits (Fig. 1.1). Sometimes, these

Fig. 1.1 A person connected with their own social network waits in front of the XIX Century socialist culture house, the *Vooruit*, in Ghent, Belgium (photo by the author, 2018)

initiatives go completely unnoticed by most citizens, or they are co-opted into institutional plans to fix the inefficiencies of public administrations. Often, they are acquired by private companies to increase their profits.

In keeping with the trend of an emerging transdisciplinary and transectorial domain in which scholars, activists and professionals collaborate, this book calls them Digital Social Innovations.

1.2 Digital Social Innovation: Why Should We Be Wary?

I bumped into the social innovation debate around the end of 2017, while researching on people's engagement in urban governance through novel forms of digital participation in the city of Ghent, Belgium. Obviously, the concept of social innovation was not new to me, but during my explorations in Ghent, it kept popping up in public discussions and interviews I had with a broad variety of key informants, social activists, entrepreneurs, administrators, academics and concerned citizens. It was as if suddenly, everybody discovered the power of an idea that is as old as modern society. It took me a while to realise that I was also working myself on social innovation; and a bit longer still to realise that the object of my research could be precisely identified as the *digital* social innovation.

In my day-to-day work of following the golden veins of promising ideas in the caves of socio-spatial complexities, I often found social innovation (and its digital variation) presented as the panacea for almost all current social problems and our diffuse sense of despair and impotence—the "malaise of modernity" (Taylor, 1992). At the same time, there was something puzzling me: if social innovation, with its seemingly endless possibilities of infiltrating and transforming the collective modes of society organisation, is so pervasive and widespread, why are its effects not markedly evident? Or, are we perhaps not attentive enough to notice them?

In discussing my doubts with social entrepreneurs, digital innovators, policymakers and colleagues, I detected a constant ambivalence. Alongside the enthusiastic accounts there was a sort of latent perplexity hovering around the idea of social innovation and—most importantly—of its practices. If these practices are supposed to allow us to overcome some of the most persistent societal pitfalls (like lack of social cohesion, inequality in various forms, economic and political divides…) why should we be wary?

Here is the rub. It is not easy to make promised changes real and most social innovation initiatives are merely able to deliver minor alterations to the status quo *via* tiny arrangements. As in a sort of Punctuated Equilibria Theory, the actual effects of social innovation (both the expected and unexpected ones) manifest, at best, in a final revolutionary event, after a long sequence of unnoticed adjustments. To grasp the complete picture, rather than focusing on details (including the technical or technological details), we need a broad perspective on the whole system. And then, we need analytic tools to deconstruct the phenomenon we are observing.

The lack of such a broad and—at the same time—analytic perspective in most of current elaborations on digital social innovation (henceforth DSI) gave me a constant sense of uneasiness.

Scholarly research on DSI is still limited, disciplinarily dispersed and rather vague on the definition of its very object of analysis. DSI lacks a theoretical pedigree and is often regarded by academics as only of interest for computer scientists or ICTs companies to develop new products. As a consequence, social science literature on DSI is relatively scarce if confronted with the significant numbers of practical experimentations. We are overwhelmed by continuous digital technology innovations; however, thoughtful considerations on their nature and social, cultural and political consequences are still rare. Innovation studies have only marginally attempted at investigating the magmatic transformations of society and the public space determined by the turmoil of digital revolution. Although the digital is massively shaping diverse areas of scientific investigation—both in terms of objects and methods, DSI is still relatively new to a large part of social research.

Most of the existing studies on DSI have been advanced in management and regional economic fields that focus on the industrial or business implications of innovation. In these economy-oriented studies, which examine the regional effects and the functioning of local networks, social innovation is approached through a functionalist and positivistic perspective. Very limited attention is devoted to the critique of socio-political and cultural implications of digital innovation.

This book attempts to open a path in a still partially virgin field of research on the relationship between DSI and the city, which, as described in the following chapters, is at once a mutable and multifaceted object of analysis and a contested terrain of practices. Building upon critical urban studies and critical internet studies, this book does not offer any normative prescription on how DSI should work but rather investigates how digital social innovators actually operate in the multiple spaces of the city.

I have been motivated in my research by a number of considerations. First, existing research in innovation management, public policies and internet studies documented and commented on the overwhelming number of diversified DSI initiatives, by focusing on the technical aspects and adopting a mono-disciplinary perspective. Most cases have been documented by practitioners and consultants interested in gaining operative and hands-on knowledge of the phenomena for replicating or exploiting best practices. On the other hand, basic academic research devoted it geographically fragmented, disciplinarily dispersed and methodologically weak attention, and produced a narrow understanding to DSI as quasi-exclusively oriented towards local initiatives. The social, cultural and organisational costs associated with such lack are significant because social aspects of the urban digital governance are relevant for the whole of society.

Second, in urban studies literature, novel forms of digital technologies applied to the city government have been characterised by a dichotomy between the critical technology-optimism of smart innovation perspectives and the radical criticism advanced by the discontents of the "wisdom of the crowd" (Lanier, 2006). This hampered the possibility of carefully scrutinising not only the products but also the production processes, the producers' agency and the effects of DSI. Therefore, I suggest balancing the positive reading of digital technologies as able to contribute to the creation of "effective, efficient, sustainable, or just" (Phills et al., 2008) cities, with the evaluation of their "dark sides".

Finally, while policy scholars have already investigated the impact of digital technologies on urban government, governance processes received relatively minor attention. Nevertheless, DSI processes are clearly feeding new fluid forms of governance characterised by continuous restructuration of power and responsibility relationships between heterogeneous, multi-scalar and hybrid actors; they are reshaping traditional forms of citizen agency, participation and entrepreneurship, by skipping away from the silos-perspective of traditional multi-stakeholder models for a distributed and transactional governance approach. Notably, the *European Commission* (EC) enthusiastically supported the adoption of digital participation and innovation tools and the integration of ICTs in public governance, by including them in the policy priorities for research and innovation. The incentives offered by the EC for engaging citizens in digital participation plans, in information-gathering and decision-making procedures are powerful means to increase both public imaginary and experimentations in different European cities. These considerations motivated the choice of Europe at large as the geographical context of investigation.

The following chapters investigate how DSI is affecting urban spaces; how it is turning into a new form of political commitment; and how the interplay between digital and social technologies produce novel processes that challenge traditional governance mechanisms. Particular attention is devoted to the different social imaginaries, political visions and narratives the city is associated with; to the governance processes that DSI promotes; and to the entanglement of multiple actors in the production of both socio-technical imaginaries and the social construction of technologies.

Our exploration starts from the evidence that the dividing line between utopian and dystopian effects produced by hyperconnected-city phantasies is incredibly thin; and, while avoiding alarming Huxleyan prophecies of a cyber-controlled society, it shows how the shadowless imagery of technology-oriented city reveals a number of practical difficulties in negotiating between public and private interests. The analysis of DSI allows us to discover how the real world and the virtual one produce one another, by originating particular spatial configurations; and how the discourses, narratives and imaginaries associated with DSI are connected with political interests that nurture existing geometries of power or construct new ones (Chesbrough, 2003; Von Hippel, 2005). In light of alternative grassroots initiatives advanced by digital collectives and social innovators, the book explores the socio-political and cultural underpinnings of the revolution produced by DSI in urban governanceand the socio-technological regimes and infrastructures supporting them. Social innovation endowed with the operational capacity of digital technologies can, in fact, tackle many of the current socio-political challenges (e.g. sharing open-source solutions and material, reducing different forms of pollutions, narrowing the democratic gap, promoting social inclusion, valuing diversity and cohesion and so on). Furthermore, it can also heavily impact on the very understanding of these problems and on our capacity to advance alternatives to neoliberal development plans for the city. In going beyond the current debate on the management of innovation, this book suggests that a serious engagement with DSI processes calls for a new social contract between digital technologies producers, institutions and citizens, as prefigured by digital activists in the *Shared Digital Europe Manifesto* (Blomen et al., 2019). This last affirms that digital innovations need to be directed by human values, and thus contribute to providing decent employment conditions; tackle inequality and social injustices in cities and work towards the achievement of environmental sustainability.

The next chapter presents and discusses concepts of social innovation that acquired distinctive features in coincidence with the digital turn, and determined the emergence of DSI. It suggests that DSI can be considered as a social technology able to impact on the material and symbolic constitutions of our world; and that DSI is, in turn, a product of distinctive societies. To explore it, we need analytic tools able to critically interrogate ideals and goals of diverse techno-social practices and their performing communities. Similar practices are discussed in Chapter 3, with reference to exemplary cases of DSI realised in European cities. The urban context represents, in fact, the preferred background for DSI initiatives; however, there has been limited attention in scholarly literature for their implications on the social, cultural and political spaces of the city. Three key concepts—extensively debated by the critical geography tradition—serve, thus, as the tenets for my investigation. These include the issue of representations, i.e. the imaginaries, narratives and discourses produced by and around DSI communities; of (re)production, i.e. the DSI contribution in perpetuating society and technological relationships in the city; and of power, i.e. the entwining of socio-political issues brought about by DSI practices. Chapter 4 examines the differences between the functionalist, reformist and revolutionary conceptions of social innovation and associates them to diverse DSI practices, and to diverse imaginaries and urban political visions promoted by distinct communities of practice. In Chapter 5, DSI initiatives are considered in terms of their capacity to intervene on existing governance processes and contribute to the (re)production of society. These interventions can both work towards or against the wide-spreading neoliberal model in urban governance, because they can both perpetuate or subvert the coded set of procedures (and the un-coded routines) that make a city working. In the business of making and unmaking our daily world, DSI initiatives materialise political aims into the design, the functioning protocols and the digital infrastructures supporting them. The political relevance of digital technologies mobilised in DSI and in the social routines these generate is explored in Chapter 6. This explains how revolutionary forms of DSI are growing political claims in line with some of the most radical visions of grassroots social movements. Their contestation of digital neoliberalism derives from the complex debate on the issues of data control, ownership and trading; on digital sovereignty and citizen participation and on the social justice implications of the digital revolution.

Overall, the exploration of the relationship between the digital and the city presented in this book aims to advance our understanding of the socio-political and cultural aspects of digital innovation in urban life and to deconstruct the apparently monolithic set of practices that goes under the name of DSI—which refers to several different approaches, visions and modes of realisation at once. In so doing, the book offers a portrait of contemporary European cities as the new battlegrounds for the control of the digital space shaped by the interplay between digital capitalism and alternative resistance movements (Diagram 1.1).

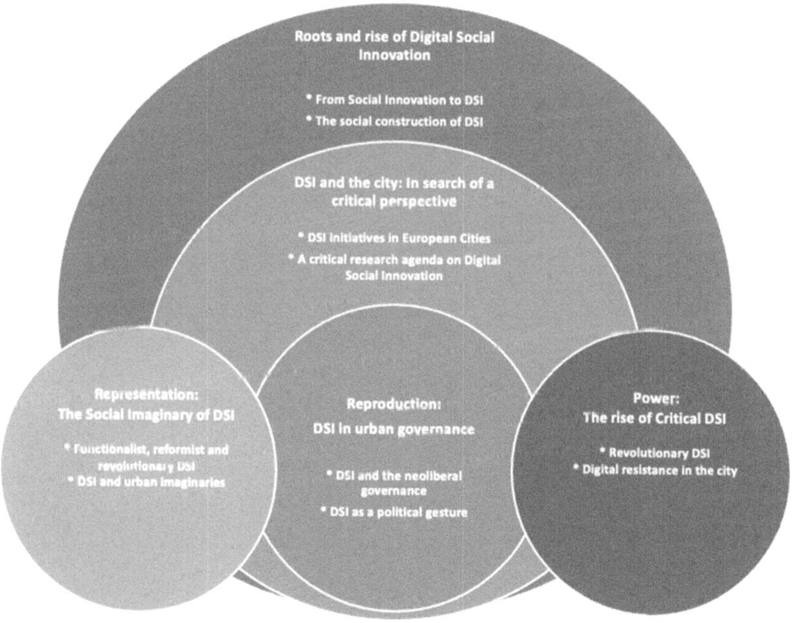

Diagram 1.1 Overall plan of the book

References

Blomen, S., Tarkiswski, A., & Keller, P. (2019). *A vision for a shared digital Europe, shared digital blog*. https://shared-digital.eu/vision/.

Chesbrough, H. (2003). *Open innovation*. HBS Press.

Lanier, J. (2006) Digital maoism: The hazards of the new online collectivism. *Edge Magazine*. Available at https://www.edge.org/conversation/jaron_lanier-digital-maoism-the-hazards-of-the-new-online-collectivism.

Phills, J. A., Jr., Deiglmeier, K., & Miller, D. T. (2008). Rediscovering social innovation. *Stanford Social Innovation Review*, *6*, 34.

Taylor, C. (1992). *The ethic of authenticity*. Harvard University Press.

Von Hippel, E. (2005). *Democratising innovation*. MIT Press.

CHAPTER 2

Roots and Rise of Digital Social Innovation

Abstract Initially interpreted as a special kind of social innovation empowered by the digital tools, Digital Social Innovation (DSI) is now recognised as an independent category of agency, endowed with peculiar features, forms and consequences. The chapter describes the roots and the evolution of the DSI concept starting from initial explorations in the social innovation domain. It clarifies how the digital turn had an impact on social innovation practices; and how the relationship between socially emancipatory innovations and the digital world represents a problematic and still underexplored field of investigation. Eventually, it explains how identifying DSI as a socially constructed social technology offers us the opportunity for a deconstructive and critical analysis upon it.

Keywords Digital Social Innovation · Social innovation · Digital turn · Social technology · Social construction of technology

This chapter builds upon an article co-authored with Filippo Corsini, titled "Digitally-enabled Social Innovation. Mapping discourses on an emergent social technology" (published on *Innovation. The European Journal of Social Science Research*, 2021, https://doi.org/10.1080/13511610.2021.1937069); and an article by the author "Digital Social Innovation and Urban Space. A critical geography agenda" published on *Urban Planning*, 2020, 5/4, 8–19.

C. Certomà, *Digital Social Innovation*,
https://doi.org/10.1007/978-3-030-80451-0_2

2.1 Introduction

In recent years, an increasing number of digital activists and digitally active citizens are tackling major social, environmental, and economic challenges by deploying the ICTs' potential. In educational provision and job seeking, political participation and democracy, scientific research, economy and business support, housing and public service improvement, they are elaborating—or just fostering the diffusion of—a plethora of new initiatives. Social mappers, hackers, open-access/source/data managers, open software (co)creators, temporary-user activists, citizen scientists, living-lab or fab-lab participants, p2p sharers and digital commoners, crowdsourcers and crowdfunders, and many others engage in the development of new techno-social solutions, including their design, prototyping, experimentation, implementation and scaling up. New products, services or processes do not only add to existing traditional solutions but reconfigure social relationships, access possibilities and local development pathways—often with scarce attention for their potential effect on the society at large. Animated by a generally progressive (despite vague) connotation, which attracts a wide consensus, these innovators promote co-creation and knowledge-based solutions that are intended to address a broad range of social needs with the proclaimed goal of contributing to a systemic transition towards socio-environmental sustainability and justice (Mulgan, 2007). These kinds of social innovation initiatives, bootstrapped by the potential of digital technological advancements, have been defined in the European context as "digital" social innovation (Bria, 2014; Caulier-Grice et al., 2012; Henning & Hess, 2010). A growing number of institutional, grassroots and business-promoted DSI projects, targeting collective socio-environmental problems that impact people's lives, have been extensively supported by private (e.g. large ICT companies, as *Microsoft* or *IBM*) and public institutions' funds (e.g. the EC).

Digital technologies (working as facilitator, enabler or transformational means) stimulated the interest of policymakers and administrators which promoted and piloted DSI initiatives in local or macro-regional contexts. This leads to the creation of rich datasets upon which policy reports build up the European position on the future for the internet in society. Despite being largely experimented and investigated *via* case-based research projects which offered consultancy on the most effective, economically profitable and efficient exploitation strategies, DSI is still a juvenile phenomenon with little scholarly research and investigation into its social, political and cultural significance. Apart from some valuable and sporadic attempts (Ayob et al., 2016; Mulgan, 2006; Murray et al., 2010), the scientific community has devoted comparatively limited attention to DSI *per se* and rather interpreted it in the broader tradition of social innovation (henceforth SI). However, the pervasive adoption of digital tools in society, the routinisation of digitally enabled processes in public life, the variety and relevance of DSI, further than the multiplicity of critical implications this entails, motivates the analyses proposed in the following chapters. In consideration of the lack of a dedicated research field, the following pages root DSI into the tradition of SI, still aware that this only partially accounts for the prerogatives of DSI, or its disruptive and questionable effects in terms of public governance and society constitution. In SI literature the digital dimension is mainly discussed as an addition to existing SI processes, with no extensive consideration of the radical transformations induced by the digital turn. While building on it, the following pages, thus, mark the differences and establish DSI as an independent object of analysis, endowed with particular features, forms and consequences that distinguish it from SI and also from the simple digital innovation investigated by computer scientists.

2.2 Social Innovation

For two centuries, social innovation has represented a vibrant field of practices and yet it is only in the last twenty years that it has seen an important increase in academic consideration (Westley et al., 2017) (see TextBox 2.1). This interest aimed to identify common ground for analysis (Mulgan, 2006) and to offer a systematic review of its multiple and often vague definitions (Caulier-Grice et al., 2012; Cajaiba-Santana, 2014; Dawson & Daniel, 2010; Moulart et al., 2013). Vagueness is an intrinsic character of social innovation, which is alternatively described as encompassing a new initiatives to promote the adoption of tools that answer social challenges, or change the role and respective positions of social actors (TEPSIE, 2014); new models of governance, empowerment and capacity building (Caulier-Grice et al., 2012); or institutional and economic transformations (Mulgan, 2007) that increase social value (Pol & Ville, 2009). In general, social innovation has always been and still is an evocative metaphor for socio-technical changes (Howaldt & Schwarz, 2010) that intervene where government policy and market solutions proved their inadequacy to cope with social needs or, even worse, where market failures have been paid for by states and civil society (Howaldt & Schwarz, 2010; Murray et al., 2010).

TextBox 2.1: The Theory of Social Innovation

The conceptualization of social innovation dates back to the Nineteenth Century, in France, when it was originally intended to criticise the naivety of early utopian-socialists and pointing out the structural constraints of capitalist dynamics (Moulaert & Van Dyck, 2013). In *Social Innovators and Their Schemes*, William Sargant (1858) explains SI emerges in periods of crisis, when generalised discontent produces a social turmoil that prefigures a transition from one to another organisational form. When the desire for change becomes strong among people, social innovators and reformers (here intended as educated and sensible individuals) give voice to the collective hopes and desires of improved living conditions. Following a first "subversive" phase characterised by the fracture between emergent social needs of a vast majority of the population and the scarcity of means to satisfy them, the concept of social innovation was adopted by sociology theories and associated with technological evolution (Westley et al., 2017). It was described by sociologist, Lester Ward (1903), as one of the pillars of social change, which combines the elements of social organisation to generate novel orders and structures. Since the early Twentieth Century, the concept of innovation entered the economy and market domain, holding strong connections with technological invention (Ogburn & Gilfillan, 1933). In this period, the idea of SI was integrated into innovation studies, notably offspringing from Joseph Schumpter's *Theory of economic development* (Schumpeter, 1934) who claimed innovation offers the impetus for the permanent process of "creative destruction" which characterises economic development. At the end of the Second World War, the capacity of social innovation theory to understand and respond to emerging social needs (in terms of care, education, housing, but also freedom of movement, social advancement, etc.) was deemed as useful to plan the extension of the organisational structures of the State and for creating new specialised institutions provided with appropriate technological tools to operate on society. This perspective fuelled the strand of sociological thought which combines social and technological change and, from the 1950s, affirmed in Science, Technology and Society studies. It was only in the '60s that the concept became a label for the methodical quest for radical, collective actions alternative to institutional ones, led by the rediscovery of politics of everyday life and small-scale solutions (Fairweather, 1967). In urban studies, this is still the prominent understanding of social innovation (Chambon et al., 1982). Scholarly reflection on social innovation exponentially increased during the two following decades (Westley et al., 2017) in the context of post-war welfare systems, in which the generation of social value was entrusted to the collaboration between the State and private initiatives (notably non-profit organisations); and in the midst of the economic boom, technological innovation was reputed as the only driver of social development. This condition has been mainly fuelled by the progressive inclusion of social innovation research into management and regional economic studies which currently represents the majority of contemporary social innovation studies.

The take-off phase, from the end of the 1980s to the beginning of the 2000s, testified a growth of research on social innovation in connection with the interest for technological innovation (van der Have & Rubalcaba, 2016); and, subsequently, although social innovation not referring to technological innovation exists, these have been largely ignored (Howaldt & Schwarz, 2010). From 2009, the economic and financial crises that unveiled the weaknesses of neoliberal systems and the fallacies of existing institutions also affected the mainstream understanding of social innovation. In fact, as Thompson notes (2019) the socially emancipatory narrative left out its less altruistic aspects that are inextricably linked to the idea and the practices of social innovation: "While the literature suggests social innovation begins in reaction to desperate situations [and] needs being left unmet by the state or market, these political origins indicate the missing complementary part in the dialectic: proactive and playful visions and imaginaries of how the world could be different. By the twenty-first century, the concept had become bound up with entrepreneurialism, increasingly embraced by governments, agencies and think tanks as a policy panacea for market failure and public sector reform" (Thompson, 2019, p. 1174). The entrepreneurial approach to social innovation has been largely advanced in innovation management studies (Drejer, 2004; Gallouj & Weinstein, 1997; Windrum et al., 2016). This also brought about some theoretical fragilities that have significant repercussions in terms of practices. Particularly, this approach regards social innovation as an instrument of neoliberalist regimes and proposes an overly simplified, a-historical and a-geographical understanding of it, while social innovation is a concept endowed with a long history and place-related characters (Westley et al., 2017). As a consequence, in the present liberal capitalism horizon (Busacca, 2013) social innovation progressively appeared in official documents of private networks and foundations, and it also entered government bodies' agenda (e.g. the European Commission's *Social Innovation Initiative*, the United States *Office of Social Innovation and Civic Participation*, the OECD *Local Economic and Employment Development forum on Social Innovations*) with the aim to influencing welfare systems. Notably, the concept gained particular traction in European local policies since its revival in the European policy agenda promoted by the *Innovation Union Flagship Initiative* (BEPA, 2010, 2014).

The definition of social innovation is not univocal (Lawrence et al., 2014); however, in general, it is described as "the creation and implementation of new solutions to social problems, with the benefits of these solutions shared beyond the confines of the innovators" (Tracey & Stott, 2017, p. 52). A review conducted by the researchers of the EC-funded project titled *The theoretical, empirical and policy foundations for building social innovation in Europe* (TEPSIE) proposed a categorization of social innovation initiatives in terms of provision of new services or products (e.g. the opening of neighbourhood nurseries and neighbourhood gardens, programmes that grant microcredit, technologies that help people with disabilities); of new practices (e.g. the creation of restorative justice and community courts, the opening of a fair trade node or a time-bank); of new rules and regulations (e.g. the application of zero-carbon policies or the incentives for energy communities); of new organisational forms (e.g. the just in time models for social challenges).

The existence of a large and magmatic variety of invention did not help in refining the ambiguous, heterogeneous and fragmented definition of social innovation (Cajaiba-Santana, 2014). From the institutional side, for instance, Murray et al.'s (2010) definition of SI has been generally adopted, as "a set of new ideas (products, services and models) that meet social needs differently from the existing one and at the same time, they create new social relationships and collaborations" (p. 3). From the grassroots movements' side, social innovation is conceptualised to be "addressing unmet social needs ('content' dimension), requiring 'changes in social relations, especially with regard to governance' ('process'), which together—if these changes are to remain durable over time—lead to new institutional configurations ('empowerment') (Moulaert et al.,

2005, p. 1976). This strand positions social innovation as driven fundamentally by social movements in civil society pursuing experiments in self-organisation, often in direct opposition to dominant institutional logics of state bureaucracy and market capitalism" (Thompson, 2019, p. 1175). The latter understanding focuses on the socially progressive aims of social innovation to "shift resource and authority flows, [and to change] social routines and cultural values of the social system that created the problem in the first place" (Westley et al., 2017, p. 4); and it stresses the novel aspects of solutions that are "more effective, efficient, sustainable, or just than existing solutions and for which the value created accrues primarily to society as a whole rather than private individuals" (Phills et al., 2008, p. 36).

Overall, characterising traits of social innovation have been identified in the novelty of both the outcome, the methods and the process (Deakins & Freel, 2009), the scaling up of innovation supported by the convergent interests of different actors and the continuous adaptation to the context of the application (Mulgan 2006, 2007). The roles of different actors are articulated in the debate on the meaning of SI between different disciplines (SI-DRIVE, 2017). The sociological reading of social innovation defines it as "a new combination and/or new configuration of social practices in certain areas of action of social contexts prompted by certain actors or constellations of actors in an intentionally targeted manner with the goal of better satisfying or answering needs and problems than is possible on the basis of established practices" (Howaldt & Schwarz, 2010). It focuses on the "practices created from collective, intentional, and goal-oriented actions aimed at promoting social change through the reconfiguration of how social goals are accomplished" (Cajaiba-Santana, 2014). On the contrary, economists (Pol & Ville, 2009) interpreted SI with reference to organisational changes in both public or private institutions and the capacity to produce economically relevant outcomes (Ruiz Viñals & Parra Rodríguez, 2013).

SI scholars pointed out that, despite that the process through which social innovation manifests vary a great deal (Caulier-Grice et al., 2012), it necessarily "involves changes to concepts and mindsets as well as to economic flows: systems only change when people think and see in new ways. It involves changes to power, replacing prior power holders with new ones" (Murray et al., 2010, p. 6). Therefore, social innovation is inherently about changing the way things are done, and systemic change can be considered as the final goal of every social innovation process—even when the degree of technical innovation introduced is low. These considerations suggest that the crucial trait that characterises SI processes resides in the aims of the undertaken initiatives (the same applies to DSI as we will see in the next chapters). The UK-based research institute NESTA defines SI as: "innovation that is explicitly for the social and public good. It is innovation inspired by the desire to meet social needs which can be neglected by traditional forms of private market provision and which have often been poorly served or unresolved by services organised by the State. Social innovation can take place inside or outside of public services. It can be developed by the public, private or third sectors, or users and communities—but equally, some innovation developed by these sectors does not qualify as social innovation because it does not directly address major social challenges" (Harris & Albury, 2009, p. 16). While, in fact, some authors consider that every innovation can be regarded as social as far as this contributes to the improvement of people's lives (Callon, 2007; Freeman & McVea, 2001), others claim that an innovation is social when it entails the participation of various actors in the transformation of knowledge, infrastructure, skills, trust and personal relationships (Sharra & Nyssens, 2019). In order to be social, thus, an innovation needs to be social both in means (i.e. performed through the concerted participation of different actors) and in ends (i.e. serves the general benefit of society) because "we define social innovations as new ideas (products, services and models) that simultaneously meet social needs and create new social relationships or collaborations. In other words, they are innovations that are both good for society and enhance society's capacity to act" (Murray et al., 2010, p. 3). Novelty is obviously a crucial element, but this novelty does not necessarily translate into something new to the world; it suffices that it is new to the users or to the context of application or brings a novel method (Mulgan, 2007) that specifically and unambiguously addresses a collective social problem (i.e. a situation which has negative impacts on people's lives and well-being), or that realises a social ideal or aspiration for a different society (more egalitarian, more sustainable and similar) (Léveques, 2001). Necessity or aspiration are thus the drivers of social innovation (Nicholls et al., 2015). In their pursuit, a social innovation will predominantly benefit a given kind of stakeholder, e.g. "individuals (abused women, alcoholics, young offenders, etc.), organisations (firms, non-profits, schools, government agencies, etc.) and territories (neighbourhoods, cities, regions, etc.)" (Sharra & Nyssens, 2019, p. 10). The optimistic expectation that the benefits of these changes will be shared equally across society is doomed to be unfulfilled. Not everybody equally benefits from the outcomes of a successful SI and some groups could even be damaged. The management studies, for instance, often stressed the centrality of social entrepreneurship as the key trait of successful social innovation initiatives, they repute to be able to reconciliate the public and the private.

2.3 Social Innovation and the Digital Turn

Since the '70s, the digital turn has induced a rapid and growing adoption of digital technologies in all the domains of social interaction (Kitchin, 2014; Leitheiser & Follmann, 2019; Westera 2012). "Digital" is a ubiquitous term with a multiplicity of meanings. It refers to the technologies characterised by binary computing architectures, but it also designates the genre of socio-techno-cultural artefacts that operate through them, and the practices of everyday life that result from our adoption of digital mediums; it names the logics and rationalities that both structure these practices and their effects; and qualifies the discourses which promote, enable and secure the increasing reach of the relevant technologies (Ash et al., 2018).

Digital devices have become ordinary presences in everybody's lives. Their adoption can be functional for both reinforcing the status quo and feeding critical considerations on it or on technologies themselves. As we will see in the following chapters, the critical analysis of digital technologies (and associated devices) requires deconstructing the material and discursive practices adopted in their creation, diffusion, and in the normalisation of their use through which they "serve and sustain particular kinds of interests [...] in society, consolidating and channelling the exercise of power" (Ash et al., 2018, p. 37).

Social innovation practices have been significantly impacted by the digital turn because new technologies have impressively decreased the communication, outreach, and access barriers. The diffusion of interconnected technologies in the last twenty years not only materially transformed our societies but also the nature of social issues and the way social innovation is carried out. As a consequence, scholars agree that "Given the fact that the pace of technology development is accelerating in power and reach, as well as continuing to reduce dramatically in cost and ease of use, we can confidentially expect the impact of digital technology on, and use by, social innovation will also strengthen considerably" (Millard & Carpenter, 2014, p. 16).

The encounter between the rapidly evolving domain of digital technologies and the social innovation domain is not frictionless. The first problem is that technology adoption is strongly lead by economic and market interests (Howaldt & Schwarz, 2010), while the ideal aspiration to solve social problems does not (necessarily) generate a profit. However, the innovation management sector, which provides the dominant narrative on SI, describes the mutual benefits of this encounter. As Matthew Thompson reminds us in a recent paper: "There is a growing sense among [...] commentators that what is needed to unleash the power of fast-evolving peer-to-peer digital technologies—and their transformative potential to circumvent mediating hierarchies of state-capitalist institutions and connect people in new democratic postcapitalist forms—are social innovations that reorganise social relations and the institutional configurations through which technologies flow and are in turn shaped" (Thompson, 2019, p. 1169). Accordingly, digital technologies have been often conceived as able to support the existing SI initiatives by enabling new types of interventions that combine the agency of off and online communities (particularly when dispersed and large-scale collective intelligence is needed in bottom-up and decentralised forms of collaboration), and transform existing processes, roles and relationships with strengthening the intermediary civil organisations (TEPSIE, 2014). For instance, some enthusiastic readings of the digital turn in social innovation suggested that "the use of so-called 'big data' to collect and analyse data [on] what social needs are being experienced by which people in different places at different times [...] can also open new perspectives for locally manufactured and very cheap products for people who otherwise have no chance of being helped" (Millard & Carpenter, 2014, p. 4). For this to happen, the affordability of effective ICT infrastructures that reduce activity costs; a generalised digital education and literacy among the population, as well as attention from the designer side to the creation of user-friendly interfaces; and the possibility for protecting data and intellectual property rights for the whole quality and responsibility chain are necessary. Nevertheless, these readings provide a quasi-exclusively economic understanding of the link between social innovation and digital technologies—being these mobilised in the neoliberal economy context or in the more limited circuits of "collaborative economy" including "sharing and redesigning goods through decentralised networks, digital fabrication, open-source and peer-to-peer technologies and hackerspaces and makerspaces" (Thompson, 2019, p. 1174).

As we will see in the following chapters, the connections between socially emancipatory innovations and the digital world extending beyond economic implications or mere processing aspects (including circulation, coding, control and ownership of infrastructures and data) represent a vast, problematic, and still underexplored domain of investigation (Amoore & Piotukh, 2015; Leszczynski, 2016). Research on social innovation only very recently acknowledged that the digital turn, further than equipping traditional social innovation processes with more powerful communication and organisational tools, determined a radical reconceptualisation of tackled problems, the choice of which problems are worthy of attention, the construction of knowledge about them and their ethical impact (Lawrence et al., 2014). Such recognition has recently opened the way for conceptualising DSI. This last refers to the peculiar capacity of digital technologies to infiltrate society and change the ontology of socio-technological assemblages; the way in which knowledge is produced and science is elaborated; and the political forms, the ends and the processes digital technologies are entangled with.

2.4 Digital Social Innovation

While often presented as a particular case of social innovation boosted by digital technologies, DSI actually constitutes an independent domain endowed with distinctive characteristics (Maglavera et al., 2019; Ozman & Gossart, 2019; Rodrigo et al., 2019; Stokes, 2020). Most of the existing scientific and grey literature defines DSI as a form of SI triggered, empowered, mediated, or even transformed by the use of digital technologies. Existing litearature exemplifies this with plenty of initiatives using social media and web connectivity to support, complement or speed existing social innovation processes. Still, the category of DSI includes innovation initiatives which would not have existed without digitally connected devices in general, and the internet in particular. These innovations change the form and functioning of society specifically *in* or *through* the digital.

An initial definition of DSI has been provided by scholars working on the EU project *Digital Social Innovation for Europe* (DSI4EU) in 2014, as a "type of social and collaborative innovation in which innovators, users and communities collaborate using digital technologies to co-create knowledge and solutions for a wide range of social needs and at a scale and speed that was unimaginable before the rise of the Internet" (Bria, 2014, p. 9). More recently it has been described as "an effective organizational network model leveraged by information and communication systems (ICTs) whose ends and means are social" (Rodrigo et al., 2019, p. 64), whose applications "increase the capacity of civic society to formulate a problem, bring it to the fore of public arenas, and engage a variety of stakeholders to jointly frame and solve this problem" (Ozman & Gossart, 2019, p. 3). Despite what might seem today as the norm, the novelty introduced by DSI, which justifies its treatment as a distinct domain of investigation, resides in the transformation of our own perspective and *modus operandi*. When we face a complex social challenge, we seek for solutions primarily by exploiting the potential of the social web, even when solutions require scientific or technical competences. In the contemporary world where the distinction between the real and the virtual seems to progressively fade away, most of people's civic engagement roots and happens in the digital space, where physical and geographical barriers are more easily overcome, and it is possible to get support and ideas, gather competences, and raise funds (Ozman, 2017). Nevertheless, the digital domain is characterised by its own logic, hierarchy, functional roles, and rules which only partially mirror those existing in the non-digital domain and consequently, need treating differently.

In general, DSI processes start from situated problems and propose solutions through co-design and co-production with affected actors. In these contexts, innovation starts from the needs, competencies, and experiences of each actor and set up new network-configured modes of interaction. Innovation in DSI not only leads to the generation of new solutions but also reconfigures the very socio-technical systems that generate them.

Although DSI initiatives differ from social innovation initiatives in terms of accomplishment and effects, the underlying motivations behind them are equally provided by social problems or ideal aspirations. For instance, *Fairphone* is a Dutch start-up adopting ethical procedures for assembling mobile phones, tracing where and how components are realised, and checking that the production, use and recycling processes associated with a device are not (too) impacting on people and planet. Another example is the *Open Knowledge Foundation*, an NGO working to provide open data for civil society organisations that want to lobby local institutions to increase their accountability and stimulate responsible behaviours. These examples suggest that civil society has a leading role in the ideation of DSI initiatives so that these can be considered as "grassroots innovation" (Smith et al., 2014) because they "use digital technologies, community engagement and collaboration, co-creation strategies and bottom-up approaches to solve societal needs, in opposition to the centralized proprietary solutions owned by a few companies" (Cangiano & Romano, 2017, p. 3546). In fact, DSI primarily emerges from the "public-private partnership based on an active role of citizens and the use of state-of-the-art information technology [...] to multiply the potential effect of grassroots initiatives" (Anania & Passani, 2014, p. 1). This character makes a difference between social innovation/DSI sector (where added value resides in social value increase) and the technological innovation sector, where success or failure of innovations is determined by its capacity to upscale and generate a profit in the marketplace (Addari & Lane, 2014).

Nevertheless, DSI initiatives are by nature multi-actorial and multi-scalar and thus success requires the involvement of institutions, research institutes and private companies, together with civil society organisations. A myriad of possible combinations of promoters, contingent and contextual conditions and regimes of technological governance produce very different DSI processes. These might include the co-creation of processes with end-users that participate in the development, adoption and implementation of the innovation (Bason, 2010; Bommert, 2010) or of new products for specific groups' needs; the transformation of stakeholders' relationships and institutional arrangements (Bates, 2012; Nicholls et al., 2015; Sörensen & Torfing, 2011); and the (re)allocation of public value outside of the efficiency and marketability of the innovation (Cels et al., 2012).

Most of the literature produced up to now on how digital technologies can be adopted in promoting social value consists of reports, position papers, proceedings and policy plans by EC specialised agencies or departments, consultants, or project leaders. The term itself emerged in the context of the *Research Development and Innovation Framework Programme* funded by the *EU Digital Agenda* policy (Anania & Passani, 2014). Up to 2014, DSI principally fell under the domain of SI; programmes supporting the creation of digital platforms (such as the *Collective Awareness Platforms for Sustainability and Social Innovation*—CAPS) revolved around grassroots experimentations and pilots with strong applicative components. Progressively, the EC promoted research and prototyping of open-source internet platforms and open hardware environments. This went together with the increased attention towards the socio-technological components of SI programmes.

Amongst others, the EU project DSI4EU was dedicated to exploring how digital technologies are mobilised to address social challenges. It identified four main areas of operation: open hardware, available for people to adapt, hack and shape tools for social change (e.g. *Safecast*, a Geiger counter with open hardware based on Arduino technology that enables citizens to measure and share radiation levels in a determined area, inspired by the need for transparent data after the incident in Fukushima Daiichi nuclear power plant, Japan, in 2011); open network, i.e. digital infrastructures that connect devices to share different kinds of resources (including goods, skills, money or time) to solve collective problems (e.g. *Guifi.net*, a mesh network using small radio transmitters that, as nodes, provide internet access to otherwise not wired areas); open data platforms, i.e. repositories that allow opening up, capturing, using, analysing and interpreting data (e.g. *OpenCorporates*, a database set up in the aftermath of the financial crisis in 2008 to make corporate data more accessible, which now includes more than 187,000 companies); and systems for open knowledge that elaborate and make available new types of knowledge (e.g. the *Open Ministry* system providing public institutions with the digital tools to involve citizens and crowdsource policy proposals for the national parliamentary vote—notably used by the Finnish Government).

All of these kinds of projects ultimately aim "to reorient technology to social ends, and to harness it to improve lives and benefit the many rather than the few; to empower citizens to take more control over their lives, and to use their collective knowledge and skills to positive effect; to make governments more accountable and transparent; to foster and promote alternatives to the dominant technological and business models—alternatives which are open and collaborative rather than closed and competitive; and to use technology to create a more environmentally sustainable society" (Stokes, 2020). Such an approach has been incorporated in the EU *Next Generation Internet Initiative* launched in 2016 by the EC, striving "to shape the future internet as an interoperable platform ecosystem that embodies the values that Europe holds dear: openness, inclusivity, transparency, privacy, cooperation, and protection of data" (EC, 2020).

2.5 The Social Construction of Digital Social Innovation

A trait of the still-nascent domain of DSI resides in the heterogeneity of the initiatives it comprises of and the multiple, sometimes divergent, perspectives it encompasses (see Chapter 4). The considerable interpretative flexibility and plural modes of performance are characteristic of technology that grows fast and randomly infiltrates different social communities, working environments and institutions. It could be argued that DSI is not a technology per se (neither does it necessarily coincide with the introduction of specific digital equipment or software), rather it is the effect of the combination of social goals with the use of digital technologies. As the following chapters illustrate, these technologies can be understood as "social technologies". Innovation scholars have amply argued that social practices are key components of innovations (Chataway et al., 2010; Hochgerner, 2013; Howaldt & Schwarz, 2010), which can, therefore, be understood as social technologies. In general, social technology is the outcome of coordinated actions and interactions directed to specific ends that becomes a standard process when associated with institutional practices and adopted by dominant social groups (van der Have & Rubalcaba, 2016). The Foucauldian interpretation proposes a more refined interpretation of social technology as a technology of societal creation and reproduction, pervaded by subterraneous mechanisms of power that make it work as a disciplinary tool able to direct peoples' behaviours and thinking (see Foucault, 1980).

Although a social technology articulates and makes use of specific technological devices (Derksen & Beaulieu, 2011), it does not identify with them or with any broader category of technological artefacts. It is rather a cultural product able to affect both the material and semiotic constitution of society. By identifying the social technologies generated and used in a determined historical period and geographical context, scholars are able to appreciate how political power is exerted through the knowledge-creation and the adoption of a wide set of techniques that act on both individual and collective mentality. A wider discussion on how DSI works as a social technology by penetrating governance processes is provided in Chapter 5.

While DSI, as a social technology, contributes to forging society and its conditions of life, in return, it is also constructed by this very society. As suggested by the Social Construction of Technology theory (Bijker et al., 2012; Williams & Edge, 1996), being a (social) technology, DSI emerges and establishes *via* the interplay of values and goals, practices, tacit knowledge and institutions. It can, therefore, be rightly regarded as the outcome of socio-political and cultural elaborations. This elaboration proceeds from initial attempts made by all interested actors to identify and conceptualise the novelty of the emerging phenomenon against an often crowded and fuzzy backdrop, which generates a condition where highly flexible interpretations overlap (Sovacool & Hess, 2017). The technology under question is associated with multiple, sometimes contrasting and inevitably variegated meanings and problems. This phase is usually followed by an un-coordinated selection process which leads to a shared cognitive framework. Divergent and context-dependent interpretations of DSI processes, elaborated by different communities of action, are discussed in Chapter 4. Different to consolidated social technologies, a consensual phase has not yet been reached on the nature, and impellent problems DSI is called to solve. As a consequence, despite DSI producing one of the most impactful forms of techno-scientific rationalities in contemporary public life, the system of reference, cognitive framework, goals, and assumptions it relies upon is not consolidated. The following chapters, thus, describe a socio-political and cultural revolution which is still in the making.

Identifying DSI as a socially constructed social technology offers the opportunity for a deconstructive and critical analysis upon it. On one hand, unpacking digital diapositives and processes mobilised in DSI makes it possible to discover how they are constituted, i.e. how the ensemble of "discourses, institutions, […] regulatory decisions, laws, administrative measures, scientific statements, [and] philosophical […] propositions" (Foucault, 1980, p. 194) designed them and made them able to circulate in society. On the other hand, it allows us to understand how DSI works as a social technology, how it produces and maintains specific social configurations. Still embryonal and tentatively conceptualised, DSI is proving, in fact, to be able to change the nature of socio-technical relationships; our understanding of how knowledge is created and enacted; and the political functioning of socio-spatial assemblages under the influx of the techno-scientific rationalities of the digital age. The understanding of DSI as a social technology is fundamental to appreciate how deeply DSI networks (constituted by innovators communities, private companies, CSOs and public sectors) are pervaded by power asymmetries; to open the black box of DSI and detect what actually drives the technological development; and to eventually overcome the agnostic assumption of a supposedly neutral DSI. These last three points are expanded upon in Chapter 6.

To disentangle the complex relationships between digital technologies and innovation for society, the following chapter proposes and explains why the adoption of the systemic view of critical geography is particularly appropriate, by starting from an exploration of the context in which DSI is performed and tracing the lines for deeper investigations.

References

Addari, F., & Lane, D. A. (2014). *Naples 2.0—A social innovation competition* (Report for Unicredit), pp. 11–12. Available at http://socialrenaissance.it/images/demo/Naples_20_UK.

Amoore, L., & Piotukh, V. (Eds.). (2015). *Algorithmic life: Calculative devices in the age of big data*. Routledge.

Anania, L., & Passani, A. (2014). A Hitchiker's guide to digital social innovation. 20th ITS Biennial Conference, Rio de Janeiro, Brazil, 30 November–3 December 2014: The Net and the Internet—Emerging Markets and Policies, International Telecommunications Society (ITS), Rio de Janeiro.

Ash, J., Kitchin, R., & Leszczynski, A. (2018). Digital turn, digital geographies? *Progress in Human Geography, 42*(1), 25–43.

Ayob, N., Teasdale, S., & Fagan, K. (2016). How social innovation 'came to be': Tracing the evolution of a contested concept. *Journal of Social Policy, 45*(4), 635–653.

Bason, C. (2010). *Leading public sector innovation: Co-creating for a better society*. University Press.

Bates, S. (2012). *The social innovation imperative. Create winning products, services, and programs that solve society's most pressing challenges*. The McGraw-Hill Companies.

BEPA. (2010). *Empowering people, driving change: Social innovation in the European Union*. European Union. Available at: http://net4society.eu/_media/Social_innovation_europe.pdf.

BEPA. (2014). *Social innovation: A decade of changes*. European Union. Available https://espas.secure.europarl.europa.eu/orbis/document/social-innovation-decade-changes.

Bijker, W. E., Hughes, T. P., & Pinch, T. (2012). *The social construction of technological systems*. MIT Press.

Bommert, B. (2010). Collaborative innovation in the public sector. *International Public Management Review, 11*, 15–33.

Bria, F. (2014). Digital social innovation. DSI4EU project Interim report. Nesta—EU H2020. Available at https://waag.org/sites/waag/files/media/publicaties/dsi-report-complete-lr.pdf.

Busacca, M. (2013). Oltre la retorica della Social Innovation. Impresa Sociale 2/2013.

Cajaiba-Santana, G. (2014). Social innovation: Moving the field forward. A conceptual framework. *Technological Forecasting and Social Change, 82*, 42–51.

Callon, M. (2007). L'innovation sociale: Quand l'économie redevient politique. In J.-L. Klein & D. Harrison (Eds.), *L'innovation sociale: émergence et effets sur la transformation des sociétés* (pp. 17–42). Presses de l'Université du Québec.

Cangiano, S., & Romano, Z. (Eds.). (2017). *Digital Social Innovation toolkit*. DSI4EU, Nesta. Available at https://digitalsocial.eu/uploads/digital-social-toolkit.pdf.

Caulier-Grice, J., Davis A., Patrick R., & Norman W. (2012). *Defining Social Innovation*. TEPSIE, European Commission Brussels.

Cels, S., de Jong, J., & Nauta, F. (2012). *Agents of change: Strategy and tactics for social innovation*. Brookings Institute Press.

Chambon, J. L., Alix, D., & Devevey, J. M. (1982). *Les innovations sociales*. Presses universitaires de France.

Chataway, J., Hanlin, R., Mugwagwa, J., & Muraguri, L. (2010). Global health social technologies: Reflections on evolving theories and landscapes. *Research Policy, 39*(10), 1277–1288.

Dawson, P., & Daniel, L. (2010). Understanding social innovation: A provisional framework. *International Journal of Technology Management, 51*(1), 9–12.

Deakins, D., & Freel, M. (2009). *Entrepreneurship and small firms*. McGraw-Hill.

Derksen, M., & Beaulieu, A. (2011). Social technology. In I. Jarvie & J. Zamora-Bonilla (Eds.), *The handbook of philosophy of social science* (pp. 703–719). Sage.

Drejer, I. (2004). Identifying innovation in surveys of services: A Schumpeterian perspective. *Research Policy, 33*(3), 551–562.

EC. (2020). *Next Generation Internet 2020*. European Commission. https://www.ngi.eu/about/.

Fairweather, G. W. (1967). *Methods for experimental social innovation*. D. Wiley.

Foucault, M. (1980). The eye of power. In C. Gordon (Ed.), *Power/knowledge: Selected writings, 1972–77* (pp. 146–165). Pantheon.

Freeman, R. E., & McVea, J. (2001). A stakeholder approach to strategic management. In M. A. Hitt, R. E. Freeman, & J. S. Harrison (Eds.), *Handbook of strategic management* (pp. 189–207). Blackwell .

Gallouj, F., & Weinstein, O. (1997). Innovation in services. *Research Policy, 26*(4–5), 537–556.

Harris M., & Albury, D. (2009). *The innovation imperative. Nesta* (The lab discussion paper, March 2009).
Henning, K., & Hess, F. (2010). *Social innovation: Concepts, research field and international trends*. ZWE der TU-Dortmund.
Hochgerner, J. (2013). Social innovation and the advancement of the general concept of innovation. In C. Ruiz Viñals & C. Parra Rodríguez (Eds.), *Social innovation: New forms of organisation in knowledge-based societies* (pp. 12–28). Routledge.
Howaldt, J., & Schwarz, M. (2010). *Social innovation: Concepts, research fields and international trends*. Sozialforschungsstelle Dortmund.
Kitchin, R. (2014) The real-time city? Big data and smart urbanism. *GeoJournal, 79*(1), 1–14.
Lawrence, T. B., Dover, G., & Gallagher, B. (2014). Managing social innovation. In M. Dodgson, D. M. Gann, & N. Phillips (Eds.), *The Oxford handbook of innovation management* (pp. 316–334). Oxford University Press.
Leitheiser, S., & Follmann, A. (2019). *The social innovation—(Re)politicisation nexus: Unlocking the political in actually existing smart city campaigns? The case of SmartCity Cologne*. Urban Studies.
Leszczynski, A. (2016). Speculative futures: Cities, data, and governance beyond smart urbanism. *Environment and Planning A*. Epub ahead of print 22 May 2016.
Léveques, B. (2001). Les entreprises d'économie sociale, plus porteuses d'innovations sociales que les autres? Colloque du CQRS au Congrès de l'ACFAS, 16 May 2001.
Maglavera, T., Niavis, H., Moutsinas, G., Passani, A., & De Rosa, S. (2019). *Digital Transformation for a better society*. ChiC, European Commission. Available at https://capssi.eu/wp-content/uploads/ChiC_D5.2_Digital_Transformation_for_a_better_society_whitepaper.pdf.
Millard, J., & Carpenter, G. (2014). Digital technology in social innovation, A deliverable of the project: The theoretical, empirical and policy foundations for building social innovation in Europe (TEPSIE). European Commission 7th Framework Programme. European Commission, DG Research.
Moulaert, F., Martinelli, F., Swyngedouw, E., & Gonzalez, S. (2005). Towards alternative model(s) of local innovation. *Urban Studies, 42*, 1969–1990.
Moulaert, F., & Van Dyck, B. (2013). Framing social innovation research: A sociology of knowledge perspective. In F. Moulaert, D. MacCallum, A. Mehmood & A. Hamdouch (Eds.), *International handbook on social innovation*. Edward Elgar.
Mulgan, G. (2006). The process of social innovation. Innovations: technology, governance, globalization, *1*(2), 145–162.
Mulgan, G. (2007). *Social innovation: What it is, why it matters and how it can be accelerated*. The Young Foundation.

Murray, G., Caulier-Grice, J., & Mulgan, R. (2010). *Open book of social innovation*. The Yong Foudation/Nesta.
Nicholls, A., Simon, J., & Gabriel, M. (2015). Introduction: Dimensions of social innovation. *New frontiers in social innovation research* (pp. 1–26). Palgrave Macmillan.
Ogburn, W. F., & Gilfillan, S. C. (1933) *The influence of invention and discovery, recent social trends in the United States,* Vol. 1, p.132 (Report of the President's Research Committee on Social Trends). McGraw-Hill.
Ozman, M. (2017). *Type of Digital Social Innovation and why they are important*. blog. https://digitalsocinno.wp.imt.fr/2017/12/21/types-of-digital-social-innovations-and-why-they-are-important/.
Ozman M., & Gossart C., (2019) Digital Social Innovation: Exploring an emerging field. ISIRC 2018 Conference, 5–7 September, Heidelberg. Available at SSRN: https://ssrn.com/abstract=3434363 or https://doi.org/10.2139/ssrn.3434363.
Phills, J. A., Jr., Deiglmeier, K., & Miller, D. T. (2008). Rediscovering social innovation. *Stanford Social Innovation Review, 6*, 34.
Pol, E., & Ville, S. (2009). Social innovation: Buzz Word or enduring term? *The Journal of Socio-Economics, 38*(6), 878–885.
Rodrigo, L., Palacios, M., & Ortiz-Marcos, I. (2019). Digital Social Innovation: Analysis of the conceptualization process and definition proposal. *Direccion y Organizacion, 67*, 59–66.
Ruiz Viñals, C., & Parra Rodríguez, C. (Eds.). (2013). *Social innovation: New forms of organisation in knowledge-based societies*. Routledge.
Sargant, W. L. (1858). *Social Innovators and Their Schemes*. London: Smith, Elder and Co.
Schumpeter, J. (1934). The theory of economic development: An inquiry into profits, capital, credit, interest and the business cycle. *Harvard Economic Studies, 46* (ed.or 1911).
Sharra R., & Nyssens, M. (2019). *Social innovation: An interdisciplinary and critical review of the concept*. Available at https://mafiadoc.com/social-innovation-literature3_5c685905097c4706578b4619.html.
SI-DRIVE. (2017). *Towards a general theory and typology of social innovation*. SI-DRIVE, European Commission.
Smith, A., Fressoli, M., & Thomas, H. (2014). Grassroots innovation movements: Challenges and contributions. *Journal of Cleaner Production, 63*, 114–124.
Sörensen, E., & Torfing, J. (2011). Enhancing collaborative innovation in the public sector. *Administration & Society, 43*, 842–868.
Sovacool, B. K., & Hess, D. J. (2017). Ordering theories: Typologies and conceptual frameworks for sociotechnical change. *Social Studies of Science, 47*(5), 703–750.

Stokes, M. (2020). *What system factors help DSI to grow and thrive?* DSI4EU, Nesta. Available at https://digitalsocial.eu/blog/145/a-first-run-of-the-dsi-index.

TEPSIE. (2014). *Social innovation theory and research: A guide for researchers*. TEPSIE. European Commission.

Thompson, M. (2019). Playing with the rules of the game: Social innovation for urban transformation. *International Journal of Urban and Regional Research, 43*(6).

Tracey, P., & Stott, N. (2017). Social innovation: A window on alternative ways of organizing and innovating. *Innovation, 19*(1), 51–60.

Van der Have, R. P., & Rubalcaba, L. (2016). Social innovation research: An emerging area of innovation studies? *Research Policy, 45*(9), 1923–1935.

Ward, L. F. (1903). *Pure sociology*. Macmillan.

Westera, W. (2012). *The digital turn. How the Internet transforms our existence*. http://www.thedigitalturn.co.uk/TheDigitalTurn.pdf.

Westley, F., McGowan, K., & Tjornbo, O. (2017). *The evolution of social innovation: Building resilience through transitions*. Cheltenham: Edward Elgar Publishing.

Williams, R., & Edge, D. (1996). The social shaping of technology. *Research Policy, 25*(6), 865–899.

Windrum, P., Schartinger, D., Rubalcaba, L., Gallouj, F., & Toivonen, M. (2016). The co-creation of multiagent social innovations: A bridge between service and social innovation research. *European Journal of Innovation Management, 19*(2), 150–166.

Links

Collective Awareness Platforms for Sustainability and Social Innovation—CAPS https://ec.europa.eu/digital-single-market/en/collective-awareness-platforms-sustainability-and-social-innovation.

Digital Agenda for Europe. https://www.europarl.europa.eu/factsheets/en/sheet/64/digital-agenda-for-europe.

Digital Social Innovation for Europe—DSI4EU. https://cordis.europa.eu/project/id/688192.

Fairphone. https://www.fairphone.com/it/.

Guifi. https://guifi.net/en.

Open Corporate. https://opencorporates.com/.

Open Ministry. https://www.oministry.com/.

Open Knowledge Foundation. https://okfn.org/.

Safecast. https://safecast.org/.

SISCODE. https://siscodeproject.eu/.

TEPSIE. https://cordis.europa.eu/project/id/290771/reporting.

CHAPTER 3

Digital Social Innovation in the City: In Search of a Critical Perspective

Abstract This chapter explains why a critical geography perspective is adequate to understand Digital Social Innovation (DSI) as a social technology operating in the urban spaces. Cities represent the preferred contexts for design, experiment and implement it—as amply documented by European Union-funded projects. However, the multiple spatial dimensions involved in DSI projects and the spatialities DSI generates have still been significantly underexplored. To this end, the chapter identifies the research lines along which to perform this space-sensitive analysis. These include reflections on representation (the space of imaginaries, narratives, discourses and values created by DSI communities); reproduction (the DSI contribution in creating and perpetuating the relationship between society and technology in the city) and power (the entwining of socio-political issues brought about by DSI practices). Through such an exploration, the ambivalent character of DSI, between economic gains of technological innovation and the alternative contestation practices, become immediately manifest.

Keywords Digital Social Innovation · Urban space and spatialities · Urban representation · Reproduction · Power · European Commission

C. Certomà, *Digital Social Innovation*,
https://doi.org/10.1007/978-3-030-80451-0_3

3.1 Introduction

Over the last 40 years, research into Social Innovation (SI) has regained interest among the international scientific community (see Moulaert et al., 2005; Westley et al., 2017). This interest coincided with the digital turn, when significant proliferation and dependence upon digital tools and processes impacted SI by granting unprecedented possibilities to increase efficiency, diffusion and effectiveness of proposed interventions. The subsequent expansion of the interactive web and the Internet of Things (IoT) produced a deep transformation in all spheres of public life and digital tools are now broadly recognised as crucial enablers of SI processes (Maglavera et al., 2019). These mushrooming connections between internet activists and social innovators are stimulating social scholarly debate on whether the digital is merely improving existing SI or changing its nature, meaning and practices. Scholars and practitioners identified, thus, a set of practices that go under the name of DSI, which are transforming the forms and functioning of society thanks to the intrinsic properties of the digital. Cities clearly stand out as the contexts in which DSI initiatives are abundant because the compactness and density of urban areas, combined with a high level of online and offline connectivity, as well as the presence of massive flows of people, goods and resources offers the ideal conditions in terms of practicality, acceptability and effectiveness of innovative initiatives (JPI, 2015). Broadly documented by EC-funded projects, SI in general and DSI in particular bear a strong link with the urban contexts in which they are designed, experimented and implemented. However, despite the few early attempts to investigate the relationships between DSI and the city (see, for instance, Stokes, 2020 or SI-DRIVE, 2017), the urban is often treated as a mere backdrop rather than the very context generating the projects in question. The multiple spatial dimensions involved in DSI projects and the spatialities DSI practices generate (i.e. the relative and relational idea of space produced by social interactions) have still been significantly underexplored up to now.

So far, considerations of DSI initiatives have come from the normative perspective of mainstream, innovation management-oriented analysis. In the innovation managers' descriptions, often aimed at illustrating the positive impact of specific projects, the considerations on the urban dimension of DSI are essentially limited to how urban ecosystems trigger or hamper innovation, or how place-based innovation can benefit local development.

Their positivist treatment of space as an objectively identifiable background (provided with distinctive physical, administrative, economic and cultural attributes) is poorly able to account for the spatial complexity of DSI and to critically consider it. The lack of critical engagement in considering the implications of DSI on the social, cultural and political spaces in the city prevents an appropriate understanding of current societal transformations.

To get a better understanding of DSI as a social technology operating in mutable and plural urban spaces, this book adopts a critical geography perspective. Through this perspective, it elaborates a thorough consideration of the intrinsic spatialities of DSI practices, and of their spatial effects on the production and conception of the city and society. This space-oriented analysis enriches the understanding of the material and symbolic transformation of urban hardware and software (Dyer et al., 2017) operated through DSI initiatives. Therefore, this chapter discusses the need for a space-sensitive analysis of DSI, and along with suggestions from critical urban theory (Brenner, 2009; Marcuse et al., 2014), STS and critical internet studies (Hunsinger et al., 2019), it defines critical geography research lines along which this is performed. These lines include the representation (the space of imaginaries, narratives, discourses and values created by DSI communities); reproduction (the DSI contribution in creating and perpetuating the relationship between society and technology in the city) and power issue (the entwining of socio-political dimensions mobilised by DSI practices).

3.2 The Urban Nature of Digital Social Innovation

The recent popularity of decentralised solutions to tackle major social challenges, including some global threats such as climate change (Hall & Pfeiffer, 2013; Heynen, 2014), and the subsequent up-scaling of originally local approaches that have proved their effectiveness (for instance in reducing pollution, increasing social cohesion, shortening distribution chains of goods), explains why cities are regarded as the appropriate contexts for social innovation initiatives (Barcelona Activa, 2018).

Problems that can be locally managed and whose identification and resolution often entail the physical proximity of citizens are approached *via* relatively short command and control chains. Arguably, these direct relationships make cities the ideal context for implementing experiments in SI.

City's residents and users often share distinctive socio-cultural and political-economic characteristics (Brandsen et al., 2016). These can be easily mobilised to intervene into place-making processes. In the document *Cities as Innovation Hub*, for instance, the Junior Research Council of the EC identified the orientation towards small and diffused entrepreneurialism, job market-related creativity and the presence of a rich substrate of local skills and culture, as the key trait that make the city attractive for testing SI (JRC, 2020).

A culture-led and social needs-oriented systemic vision, in fact, provides the city with a competitive advantage over different spatial configurations (i.e. non-urban areas, regional or national aggregates) (Drydyk, 2013; Ibrahim & Alkire, 2007). The point is clearly affirmed in the EC *Strategic Research and Innovation Agenda* (JPI, 2015) whose aim is to foster the experimentation of innovative concepts, processes, tools and governance models. European cities can serve as test beds for co-designed and co-implemented solutions, which can be integrated into development plans and policies. Not surprisingly, most of the existing DSI projects worldwide, too, are realised in the cities (INEA, 2020; Vandecasteele et al., 2019).

However, apart from this general observation, scientific research has devoted very limited attention to the spatial dimensions of the city (including the physical, social and symbolic ones) entailed by DSI processes. Among these spatial dimensions, the digital one, with its material infrastructures and immaterial space of codes and connections, is acquiring an ever-increasing relevance because: "When cities are combined with digital technologies, our urban habitat becomes the most sophisticated technology for interaction ever created" (Han & Hawken, 2018, p. 2). The resulting augmented space, which characterises contemporary cities, makes them able to harness tangible and intangible resources thanks to the density of networks in which proximity to resources is substituted with proximity to knowledge (Landry, 2015).

And here comes the rub, most reports on innovation management and regional policy literature adopt quite an apologetic approach when describing the interaction of DSI with the urban context. They praise the intervention possibilities offered by digital technologies, which are supposed to endow social creativity with new operational capacity to allow experimenting with spontaneous, citizens-led, socio-ecological alternatives. In so doing, they disregard the criticalities that come along with the progressive transfer of public responsibilities and tasks to citizens' organisations and often to private companies, whose end resides in creating an economic profit, rather than in operating for the common benefit. When DSI is described and practised as a further way to outsource public services to private actors, it contributes to the advancement of the neoliberal urban agenda. Such an understanding of DSI "mirrors more recent policy obsessions with creativity—a close relation of innovation—expressed in New Labour's doomed love affair with the 'creative industries' and the rise of Richard Florida's ideas around attracting the coveted 'creative class' as a panacea to the economic problems of de-industrializing cities" (Thompson, 2019, p. 1172).

Nevertheless, as soon as we approach the issue of DSI and dig deep into its forms, meaning and aims, we realise that there are also decentralised and bottom-up initiatives (Komninos, 2009; Ratti & Townsend, 2011), advanced by digital activists and social innovators that support a radically different socio-political vision, which is equally utilising the city as the space to operate in.

And so, the ambivalent character of DSI become immediately manifest. It plays on the dividing line between the "dominance of capitalist discourse [and the] close association with strictly formal understandings of economic life as essentially competitive and individualistic" (Thompson, 2019, p. 1172), which is often associated with the gains of technological innovation; and the contestation practices which seek for an alternative to cyclical economic and financial crises that manifest along the reproduction process of the global capitalism (Evans & Karvonen, 2010).

3.3 Digital Social Innovation Initiatives in European Cities

The eminently urban nature of DSI can be easily detected by exploring the vast panorama of projects running in European cities. The number and increasing impact of this is not at all represented in the scarce number of scholarly contributions on the topic. The reason why European cities emerged as the privileged places for experimenting innovative solutions, can be found in the progressive elaboration of communitarian development strategies which significantly supported civil society initiatives in response to pressing socio-environmental challenges (notably to migration, unemployment and rising poverty, ageing or climate change) (Salamon et al., 2003; van der Have & Rubalcaba, 2016), as a response to the inefficacy of traditional instruments (Hämäläinen & Heiskala, 2007; Stern, 2007). Following the *Review of the EU Sustainable Development Strategy* the EC adopted specific policies and funding programmes for supporting pioneering ideas able to produce changes in social behavioural patterns (EC, 2005, 2011).

A large part of the expected systemic change is associated with open governance and direct democracy (EC, 2013) and leverages on people's engagement and empowerment through dedicated participatory technologies. Digital technologies transformed many areas of business, notably the whole production and service provision system, with large investments of public money to supporting them. Comparatively less attention has been to the adoption of digital technology to address social challenges, despite this being one of the more promising (and contested, see Chapter 6) areas in which joint ventures between tech entrepreneurs and public institutions are expected. Since the elaboration of the *Digital Agenda for Europe* in 2010 (EC, 2014), digital innovation has been presented as the enabling factor to reshape interaction routines between institutional, business, research and civil society actors in a democratic, effective and sustainable way (Hearn et al., 2011; Nielsen, 2006). Progressively DSI became a specified definition of SI in line with the EC policies, notably the Lisbon's Strategy for a "smart, sustainable and inclusive growth" (EC, 2010a) and the *European Strategy for Development* (Barroso, 2014; EC, 2008, 2010b). Along this mainstreaming process, the functionalist perspective on DSI permeates policymakers, consultants and even practitioners' work. The rich corpus of (mainly applied) research results on DSI have been significantly supported by EU-funding programmes (Table 3.1). Almost all of the relevant projects revolve around pilot applications of new products and services in urban or regional contexts and combine research with the experimenting of open source, distributed and collaborative tools intended for community empowerment.

Most of these projects analyse a large array of case-studies or test new products and services (e.g. *CIMULACT*; *EDFx; SISCODE*) that are aimed at life quality enhancement (e.g. *OPENCARE, WE4AHA*) or environmental care (e.g. *CAPTOR; SHELTER; HYPERION*), in different cities. Other projects provide first-hand results on the economic and social potentialities enabled *via* digital tools (e.g. SI-DRIVE; SIMPACT;

Table 3.1 Examples of the EC-funded projects (from 2013 to 2020) contributing to the establishment of DSI research and application domain

Macro-area of intervention	*Acronym*	*Title*	*Starting date*
Innovation (theory and application)	D-CENT	Decentralised Citizens ENgagement Technologies	2013
	SI-DRIVE	Social Innovation: Driving Force of Social Change	2014
	Transit EU	Transformative Social Innovation Theory	2014
	SIC	Social Innovation Community	2016
	DSI4EU	Digital Social Innovation for Europe	2016
	netCommons	Network Infrastructure as Commons	2016
	Open4Citizens	Empowering citizens to make meaningful use of open data	2016
	COMRADES	Collective Platform for Community Resilience and Social Innovation during Crises	2016
	DECODE	Decentralised Citizens Owned Data Ecosystem	2016
	MAZI	A DIY Networking Toolkit for Location-Based Collective Awareness	2016
	MAKE-IT	Understanding Collective Awareness: Platforms with the Maker Movement	2016
	EMPATIA	Enabling Multichannel Participation Through ICT Adaptations	2016
	FI-GLOBAL	FI-GLOBAL: Building and supporting a global open community of FIWARE innovators and users	2016
	Families_Share	Socializing and sharing time for work/life balance through digital and social innovation	2018
	SISCODE	Society in Innovation and Science through CODEsign	2018
	GrassrootWavelengths	Grassroot Wavelengths: Highly Networked Grassroots Community Radio through a scalable digital platform	2018
	OpenHeritage	Organizing, Promoting and Enabling Heritage Re-use through Inclusion, Technology, Access, Governance and Empowerment	2018
Economy	VirCoin2SME	Social, Complementary or Community Virtual Currencies Transfer of Knowledge to SME: a new era for competitiveness and entrepreneurship	2014
	SIMPACT	Boosting the Impact of Social Innovation in Europe through Economic Underpinnings	2014
	OpenMaker	Harnessing the power of Digital Social Platforms to shake up makers and manufacturing entrepreneurs towards a European Open Manufacturing ecosystem	2016
	PIE News	Poverty, Income, and Employment news	2016
	PROFIT	Promoting Financial Awareness and Stability	2016
	ChainReact	Making Supplier Networks Transparent, Understandable and Responsive	2016
	DISCE	Developing Inclusive & Sustainable Creative Economies	2019
	CICERONE	Creative Industries Cultural Economy Production Network	2019
Environment	POWER	Political and sOcial awareness on Water EnviRonmental challenges	2015
	SOCRATIC	SOcial CReATive IntelligenCe Platform for achieving Global Sustainability Goals	2015
	ThinkNature	Development of a multi-stakeholder dialogue platform and Think tank to promote innovation with Nature based solutions	2016
	hackAIR	Collective awareness platform for outdoor air pollution	2016
	CAPSELLA	Collective Awareness PlatformS for Environmentally-sound Land management based on data technoLogies and Agrobiodiversity	2016
	SavingFood	An innovative solution to tackle food waste through the collaborative power of ICT networks	2016
	ASSET	Instant Gratification for Collective Awareness and Sustainable Consumerism	2016

(continued)

Table 3.1 (continued)

	CAPTOR	Collective Awareness Platform for Tropospheric Ozone Pollution	2016
	STARS4ALL	A Collective Awareness Platform for Promoting Dark Skies in Europe	2016
	CROWD4ROADS	CROWD sensing and ride sharing FOR ROAD Sustainability	2016
	R2PI	Transition from linear 2 circular: policy and innovation	2016
	PTwist	PTwist: An open platform for plastics lifecycle awareness, monetization, and sustainable innovation	2018
	Pop-Machina	Collaborative production for the circular economy; a community approach	2019
	REFLOW	constRuctive mEtabolic processes For materiaL flOWs in urban and peri-urban environments across Europe	2019
	SHELTER	Sustainable Historic Environments hoListic reconstruction through Technological Enhancement and community based Resilience	2019
	ARCH	Advancing Resilience of Historic Areas against Climate-related and other Hazards	2019
	HYPERION	Development of a Decision Support System for Improved Resilience & Sustainable Reconstruction of historic areas to cope with Climate Change & Extreme Events based on Novel Sensors and Modelling Tools	2019
Health	Made4You	Open and Inclusive Healthcare for Citizens Based on Digital Fabrication	2013
	OPENCARE	Open Participatory Engagement in Collective Awareness for REdesign of Care Services	2016
	WE4AHA	Widening the support for large scale uptake of Digital Innovation for Active and Healthy Ageing	2017
	SHARE4RARE	Social media platform dedicated to rare diseases, using collective intelligence for the generation of awareness and advanced knowledge on this large group of diseases.	2018
Youths	DigiGen	The impact of technological transformations on the digital generation	2019
	ySKILLS	Youth Skills	2020
	DIGYMATEX	Establishing a comprehensive understanding and taxonomy of children's digital maturity	2020
	CORE	CORE - Children Online: Research and Evidence. A knowledge base on children and youth in the digital world	2020
Innovation Policy	CIMULACT	Citizen And Multi-Actor Consultation On Horizon2020	2015
	EDFx	European Digital Forum Thought Leadership and Policy Network Exchange	
	ChiC	Creating high impact for CAPS	2016
	IMPACT GROWTH	Impact growth: European superstars for future internet	2016
	DSISCALE	Supporting the scale and growth of Digital Social Innovation in Europe through coordination of Europe's DSI and CAPS Networks	2018
	FI-NEXT	Bringing FIWARE to the NEXT step	2016
Territorial governance	URBANA	Urban Arena for sustainable and equitable solutions	2019
	CIRCuIT	Circular Construction In Regenerative Cities (CIRCuIT)	2019
	CityLoops	Closing the loop for urban material flows	2019
	FAIRshare	FAIRshare. Farm Advisory digital Innovation tools Realised and Shared	2018
	DESIRA	Digitisation: Economic and Social Impacts in Rural Areas	2019

TransitEU). A more detailed graphic representation of the specific topics associated with the DSI (and some SI endowed with digital tools) initiatives in EC-funded projects is offered in Fig 3.1. This representation has been elaborated *via* the software-aided scraping of 1987 reports retrieved on the EU CORDIS database (the repository of EC-funded projects outcomes). Frequently mentioned topics (upper right-hand side of the

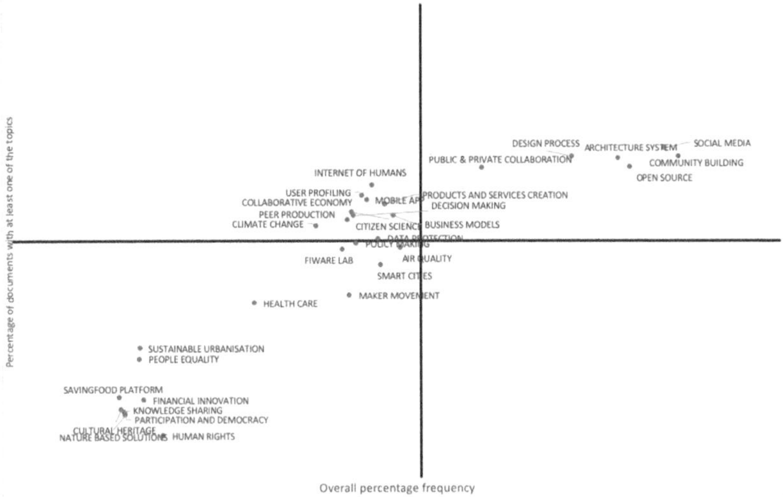

Fig. 3.1 Results of text mining conducted on the content of 1987 reports of EU projects (retrieved by using the keywords "DSI", "Collective Awareness Platforms" and "ICT-enabled Social Innovation" in *Wordstat 8*), showing the most frequently mentioned topics (elaboration by Filippo Corsini)

chart) include the different tools (e.g. social media), methodologies and approaches (e.g. open-source design process) adopted in DSI-relevant projects. Centrally the chart shows that overall, the projects pay close attention to the technology-oriented approach in order to deal with social and scientific issues (e.g. citizen science, data protection, internet of humans). On the left-hand side of the chart, less frequently mentioned topics include different aspects undertaken in some specific projects (e.g. saving food platform, health care, human rights).

Despite the numerous EC-funded DSI initiatives, the overall panorama extends much further into a flourishing array of public, private and in-between projects which continue to crowd the space of digital urban governance (a space we will explore further in Chapter 4), and introduce (sometimes fleetingly) changes in social and spatial constitutions of cities and in their functional processes. Clearly, what these processes are, should or should not be, is a contested issue, as will be explored further as we continue.

A few attempts have been made at mapping this extensive, variegated and complex panorama on a geographical base, or on the basis of adopted management processes, or addressed issues (see the EU projects SI-drive, D-Cent, TESPIE and DSI4EU). Nevertheless, the dynamicity, heterogeneity and volatile characters of many initiatives have made it almost impossible to draw an overall panorama. Indicative figures, provided by the EU DSI4EU project in 2019, suggest that the movement of digital social innovators in Europe have over 2240 collaboration projects brought forth by about 1500 organisations. In geographical terms, the projects span from the central to peripheral areas. From the Finnish *Nappy Naapuri*, to the Portuguese *WeProductise*; from the Icelandic *Citizen Foundation* to the Greek *Open Source* software society—these create spaces of relational proximity and interest linkages between local communities.

The extent of variety regarding initiatives under the name of DSI makes it difficult to provide a comprehensive account of this emerging domain. DSI initiatives are vastly diverse in terms of the people who promote and perform them (including public administrations, research institutes, different ranges of private companies, CSOs or NGOs, spin-offs and often networks involving heterogeneous social actors); in terms of management processes (such as top-down, bottom-up, grassroots or middle-out patterns) and of participant engagement (ranging from passive participation in data collecting processes to active process-leading). The purposes of DSI projects also vary a great deal but, as suggested by Francesca Bria, they propose "radically new ways of organising many of the essentials of life – from money and health to democracy and education. Its forms are still emergent, some growing very fast, others still being quite marginal. It has been almost entirely invisible to policymakers and has had none of the extensive support that has gone into digital technologies for the military, government and business. But it has the potential to contribute to three of the most important challenges facing Europe: reinventing public services, often in less costly ways; reinventing community, and how people collaborate; and reinventing business in ways that are better aligned with human needs" (Bria, 2014, p. 8). In pursuing these aims, an extensive number of sub-issues are tackled through independently promoted DSI projects, which root in or apply to European cities and the wider regions. Most existing DSI projects focus on offering educational opportunities for people who cannot attend existing institutions, or providing the means to expand one's own cultural competence and skills in both traditional and novel fields, with dedicated curricula and virtual classes; on inventing and upscaling new models for public participation in policy-discussion and decision-making from simple opinion gathering to more articulated e-voting systems; on enriching scholarly scientific research with the contribution of citizens scientists or fostering technology sharing processes and DIY components for makers and fab-labbers; on creating new venues where online and offline solidarity links can be activated and contribute to the inclusion of marginalised social groups, and to sharing economy and collaborative working opportunities; on helping people to monitor and improve their health conditions in cases of chronic diseases; on fostering new business models focused on the creation of social value in an ethical and sustainable economy system. Some of the most common kinds of DSI projects applied to the urban context are included in Table 3.2.

To reach their goals, these initiatives exploit the network effect, which refers to the incremental benefit gained by participants from the joining of new participants in the network. Firstly, they all take advantage of the fast connectivity and the easiness of data-gathering provided by social media and the peer-to-peer protocol, by crowdsourcing, crowdmapping processes and similar. Secondly, most of them adopt open-source technologies, online marketing and learning tools. Thirdly, some of them leverage on the IoT and Artificial Intelligence (AI) functions, big data repository, geotagging systems, GPS or GIS technologies or complex participatory platforms. While established technologies are those that can be more easily incorporated into DSI initiatives because these are already consolidated (e.g. open data, open hardware, crowdsourcing, citizen sensing and digital fabrication), newer and emerging technologies (e.g. blockchain, AI and virtual and augmented reality), which have been adopted up to now in the commercial sector, are now increasingly used in addressing social challenges in the non-commercial sector (Baker et al., 2017). Project examples include *Robot Lawyer*, which supports refugees' applications in the UK (among other places); *iris.ai*, an artificial intelligence helping research on climate change and decision-making; the *Descartes Labs*, which combines satellite imagery data to predict crop yields and shortages in different areas of the world; the *NuEyers* project, which adopts AR and VR in disease diagnosis and treatment and the *AID:Tech* project, which uses blockchain technology for digital identity verification of financially excluded people.

Table 3.2 A selection of DSI initiatives in urban contexts promoted by local institutions, civil society, social enterprises and associations in Europe

Category of DSI initiative	*Examples*
Collaborative governance and planning processes	***Better Reykjavik*** (Island). An online consultation forum used by local administration to consult citizens and collaborative urban planning ***Liquid Democracy***, Berlin (Germany). An organisation that promotes civic participation by providing an open-source software, Adhocracy, and other digital tools ***Hackney CAB Crowd Map***, London (UK). An interactive platform to map housing benefit shortfalls and threatened evictions across the borough ***Madame la maire, j'ai une idée***, Paris (France). A portal for the city administration to promote collaborative campaign and support citizens' promoted projects
	Writetothem, London (UK). A website (using *mySociety* software MapIt) that provides an easy way to contact councillors and elected representatives
Participation in democracy processes	***Rahvakogu***, Riga (Estonia). A one-off initiative in 2013 in which citizens discussed political corruption; it resulted in an ongoing project, *CrowdLaw*, which works as a permanent assembly to discuss and deliberate on legislation ***Decide Madrid*** *(*Spain). A platform for public decision making on new local laws, participatory budgeting and new proposal discussion ***Open Ministry***, Helsinki (Finland). A non-profit organization providing citizens and administration the tools for *crowdsourcing legislation initiatives*
Strengthening community cohesion and solidarity links	***StreetLink***, several cities in England and Wales (UK). A platform enabling members of the public to connect people sleeping rough with the local services that can support them ***Peerby***, initially Amsterdam and Rotterdam (Netherlands), now global. Sharing platform that facilitates free or paid-for transactions between neighbours of products ***Nappi Naapuri***, Helsinki (Finland). An association bringing people together for getting help, company or advice at the local level ***Refugeeswork***, Berlin (Germany). A project helping refugees and newcomers to build digital skills and to find freelance work in the hosting country Several crowd funding platforms, generally nation-based, e.g. ***Buonacausa*** *(Italy)*, ***Spacehive*** *(UK)*, raising money for projects that help in local areas
Knowledge and skills sharing and science production	***Hack the City***, Sheffield (UK). An online initiative that supports hands-on event to use public data to test and prototype ideas, apps and products that help using open data for a better urban and civic experience ***Airportalen***, Alnarp, Umeå and Uppsala (Sweden). A website by The Swedish University of Agricultural Sciences for reporting and collecting observations of the plants, animals and fungi of Sweden ***Mapping for Change***, London (UK). An organisation offering a range of participatory mapping services ***Hackerspace*** Lublin (Poland). A local space for scientists, engineers, and artists to engage with local culture and co-creative repairing initiatives ***Aquapioneers***, Barcelona (Spain). A company that provides open-source hydroponic solutions (including design and installation)
Digital Technological accessibility	***Arduino***, initially in Ivrea (Italy), now global. An open-source hardware and software company, and project and users' community designing and producing single-board microcontrollers and microcontroller kits for building digital devices ***WeProductise***, Ponte de Lima (Portugal). A graphics and communication company aimed to create Sustainable, Inclusive and Positive Design ***FABLAB Manorhamilton*** (Ireland). A digital fabrication facility lab for enterprises, creatives, schools and hobbyists' access

(continued)

Table 3.2 (continued)

Social value-based production and working models	***Commons Network,*** initially in Berlin, Brussels, Amsterdam and Valencia, now all Europe. A civil society organisation supporting commoners and defend the commons by network-building, consultancy, campaigning, and lobbying ***Future everything,*** initially Manchester (UK), now Europe. An international community of artists, technologists and industry collaborators promoting conversation and event promotion on the future of digital culture ***ThingsCon,*** initially Netherlands and Germany, now global. An initiative that explores and promote the development of fair, responsible, and human-centric technologies for the IoT ***Waag,*** Amsterdam (Netherlands). An organisation of research and grassroots groups and institutional partners promoting DIY in hardware, production processes and materials
Transparency and information disclosure	***K-Monitor,*** Budapest (Hungary). An organisation developing technology solutions to foster transparency and participation on issues related to freedom of information, political finance, public procurement, and local governance ***Open Data Aarhus*** (Denmark). A platform opening urban data to citizens ***Environmental Agency DataShare,*** Leeds (UK). An organisation that makes environmental information available to deliver sustainable environmental improvements
	Swirrl, Stirling (UK). A company providing government organisations with_**Publish MyData** software platform that make data more accessible and re-usable by citizens

3.4 The City as a Background for Digital Social Innovation

As mentioned before, most of the project reports and scientific research investigating DSI initiatives in urban contexts treat the city as an inconsequential backdrop and devote a limited attention to the multiple spatial dimensions of DSI, beyond mere physical spatiality to socio-cultural, material-semiotic and political ones. Mainstream contributions elaborate from innovation management and regional studies (Rissola et al., 2017). Therefore, up to now, principal streams of research on the spatial dimensions of SI and DSI focused on the analysis of conditions that allow these initiatives to emerge and to bring successful production processes and on the economic impacts they generate on interested territories (Cajaiba-Santana, 2014; Dawson & Daniel, 2010). In this way, the urban context is regarded as the ecosystem where digital innovation can flourish through a harmonic orchestration of ideas, institutions, regulations and policies (Bria, 2015; Granstrand & Holgersson, 2020; Whittle et al., 2012; EC, 2020). For instance, the collaboration between the municipality and the privately funded *Technology Park* in Ljubljana which hosts start-ups, co-working spaces, geek houses and hackathons, is proposed by the *EU Smart Specialisation Platform* as a good example of such an ecosystem (Bučar & Rissola, 2018; EU Science Hub, 2020). It is well recognised that bringing forward a successful innovation process is not a matter of luck but of enabling conditions that support it (Boelman & Heales, 2015). The place-specific conditions that impact on the inventive ability and operational capability of digital social innovators give place to a sort of "place-baseness" (Millard & Carpenter, 2014; Eckhardt et al., 2016). As Millard and Carpenter noted, "disruptive innovations that come out of nowhere are very rare [...] most are incremental changes built on the underpinnings of other knowledge, technologies or platforms. What is important for most innovations to occur is a set of enabling conditions that trigger people and groups with the right knowledge and skills to recognise (even serendipitously) an incremental step that can be taken at that moment in time" (2014, p. 15). Secondly, the impact of innovation on local development is generally assessed on the basis of managerial or economic effects. Particularly, with regard to impact, DSI initiatives are praised when they "facilitate civic engagement not only by making it easier, cheaper and less time-consuming but also [by expanding] the space of civic engagement by bringing new ways of engagement" (Ozman & Gossart, 2019, p. 29). This is probably due to the idea that the mobilisation of digital tools is able to bring about institutional or organisational changes in local institutions' routines and working culture (Alvord et al., 2004; Mulgan, 2006; Seyfang & Smith, 2007; van der Have & Rubalcaba, 2016;). Depending on the degree of technological readiness, on the regime of technological governance and the proneness to innovation, in fact, DSI initiatives can be realised to test new products on target groups (Mair & Ganly, 2010); engage end-users in co-creation processes (Bason, 2010; Bommert, 2010); changing stakeholders' relationships and institutional arrangements (Bates, 2012; Nicholls et al., 2015; Simon et al., 2017; Sörensen & Torfing, 2011) or even re-allocate the value of innovation outside of its marketability (Cels et al., 2012).

In the last ten years, the tendency to reduce SI to the creation of social enterprises, and their contextual conditions to economical determinants has been mitigated by urban scholars' interest for initiatives that proliferate outside institutional settings, experimenting with new social relations and ways of governance (MacCallum et al., 2009). This move produced a first distancing from the mainstream neoliberal understandings of innovation towards a socio-political transformative interpretation (Moulaert & Van Dyck, 2013). SI, from a strategy to satisfy certain individual and collective needs, turned into an opportunity to strengthen solidarity in social relations and to vehiculate collective capacity of group-building. Territories and local places where this happens has been analysed through new lenses, for instance, the approach of Integrated Area Development (MacCallum et al., 2009). It observes the city as composed of various, interrelated socio-ecological dimensions, whose relationships can be fuelled and improved *via* new structured governance systems based on the experiences of civil society proposed by SI initiatives. Among others, the EC project SINGOCOM analysed the territorial dimensions of SI (Drewe et al., 2008; Oosterlynck et al., 2013) and proposed the adoption of a non-institutional model of innovation (Moulaert et al., 2005) where SI is not only conceived as an attempt to answer people's deprivation in terms of economic needs (e.g. housing, employment, etc.), but also of cultural (e.g. self-expression, creativity) and political ones (e.g. participatory citizenship, self-government) (Thompson, 2019).

This perspective opened a way to investigate social innovation, and DSI, outside the mainstream of innovation management perspective, and with more attentive consideration for socio-spatial determinants. This inspired research on the influential factors that fuel the emergence of DSI, for instance, in EU projects DSI4EU and SI-DRIVE. Together with more expectable factors (including, for instance, the existence of dedicated policy measures, resources availability, active citizenry, attractiveness for the creative class, presence of high-level research and education institutions), they suggested some distinctive traits that characterise cities where digitally enabled social innovators operate. Notably, these include the presence of social problems that both the public and the market have failed to address (Murray et al., 2010) and of a civil society provided with sufficient technological literacy and technology access (SI-DRIVE, 2017). According to the *City Index Report* elaborated by the DSI4EU project's researchers (Baker et al., 2017), this last is a critical element that allows the emergence of DSI initiatives, together with dedicated local and national policy initiatives, and the availability of good quality and accurate data. However, they have not been able to assess the impact of geographical and contextual differences on the emergence of DSI processes. The existence of favourable conditions proved to be necessary but not sufficient. The DSI4EU team concluded their report by stating that it is "simplistic to think [that] the eco-systemic factors and activity should correlate closely [...] Many DSI initiatives emerge as a response to unfavourable social, political and economic contexts, in an attempt to address social issues that have been overlooked by traditional institutions.

[…] Some of the cities which are high-ranking in the index might simply have less need for DSI; and some of the cities which are low-ranking in the index might have conditions which lead people to develop DSI against the odds" (Stokes, 2020).

Overall, the results of existing studies suggest that a different approach can better illuminate the relationship between DSI and the city. Research should operate a deconstructive analysis of the social, cultural and political spaces of the city that are involved and transformed by the mushrooming of DSI processes and the spatialities these generate (see Sect. 3.5). Following the inchoate suggestions of a different scholarly tradition, particularly that of critical geography, can lead us to overcome the idea of the city as a mere background for DSI initiatives and as a physically constrained tangle of actions-and-reactions, towards a thicker comprehension of the relationship between DSI and multiple urban spaces and spatialities.

3.5 Behind the Background: The Social, Cultural and Political Spaces of the Augmented City

Critical geography research can offer an alternative perspective for a deeper consideration of what triggers DSI processes in the city and what social, cultural and political impact this might have, beyond geographical economic ones. Following some of its investigative lines, it is possible to detect the socio-cultural geography of DSI, including the relational proximity to where tackled problems manifest or where competences to solve them are made available; the space of agency conceptualised by involved actors and the symbolic meanings they circulate through the process and the impact of DSI processes on people's recognition, empowerment, accessibility and inclusion. The scholarly tradition of critical geography, in fact, does not provide a coherent epistemology but rather an orientation of research that embraces political stances (Murdoch, 2005; Soja, 1989) and in so doing, unveils inequalities, injustices and dominations perpetuated through mentalities creation or material forms of coercion (Blomley, 2006), to advance emancipatory claims *via* engaged scientific practice. One of its key traits resides in the rejection of the positivist approach for describing reality, and in the belief that socio-spatial structures are contingent and mutable configurations created by those who deal with them. It is exactly this specific understanding of space that makes it an appropriate conceptual tool for a deeper exploration of the DSI as a city-creation practice.

However, it needs to be signalled that the topic of DSI hasn't been directly investigated by critical geographers up to now. Therefore, it is from the rich set of contributions on the socio-spatial effects of the digital revolution in the city that we can retrieve useful insights for the exploration conducted in this book.

Critical geographers extensively considered how both the concept and experience of space underwent profound changes in the digital age. The feared annihilation (Castells, 1996), compression (Harvey, 1989) or flattening of space (Cairncross, 1997) due to the digital turn and the rise of networked ICTs have been initially regarded as potentially leading to the irrelevance of geography (Friedman, 2005; Ash et al., 2018). It soon became evident that this was, and is, not the case. In the digital age, physical spatial relations are not extinguishing but rather transforming (March & Ribera-Fumaz, 2014; Cardullo & Kitchin, 2018; Vanolo, 2013). Territorial linkages based on physical proximity are increasingly complemented with connections among relationally proximate loci that are functionally interoperable. In urban contexts, the digital turn determined an expansion of the urban space, by merging material dimensions with the virtual ones, endowed with new organisational logics (de Waal, 2014), into a single reality—a sort of "hypercity" (Landi, 2019; Massey & Snyder, 2015). The virtual space is an extension of the city around, inside, beyond and behind its physical space. This "augmented urban space" (De Cindio & Aurigi, 2008) is the effect of subtle yet pervasive changes in how society keeps on doing usual things but in a new way: "There are no entirely digital lifestyles – or, if there are, we are talking about a fairly irrelevant minority – yet at the same time most lifestyles are digitally-enhanced, in many different as well as subtle ways. In the augmented city, 'virtual' and 'physical' spaces are no longer two separate dimensions, but just parts of a continuum, of a whole. The physical and the digital environment have come to define each other and concepts such as public space and 'third place', identity and knowledge, citizenship and public participation are all inevitably affected by the shaping of the reconfigured, augmented urban space" (De Cindio & Aurigi, 2008, p. 1).

It might be contested that the virtual space is made of temporary events (such as a platform, a web page, a social media profile) subject to sudden transformations or even disappearances depending on the unstable interests, alliances or conflicts of its creators. The *long durée*, the fixity and obduracy of the physical space are apparently opposed to the ephemerality of the virtual one; but, this contraposition is misleading for two reasons. First of all, physical space and its places (such as a city or a neighbourhood) are themselves contingent formations happening in a fluid (physical) space, springing out from the interplay of contrasting forces. Secondly, the virtual presence of a city in the webspace is not a duplication or reflection of the physical one, but it is a further dimension of the city itself. This latter dimension is supported by dedicated infrastructures (including broadband cables, the www domain, servers, sensors, etc.) and provides new services (such as citizen participation portals, online sharing economy platforms and similar). Not surprisingly most of the digital platforms adopted in DSI processes are tightly connected with the geographical context of realisation (e.g. time banks, sharing, e-petition or GIS-based maps) (Goodchild, 2007; Verona et al., 2006). The diffusion of digital technologies, the interactive web and the IoT has transformed the urban hardware and software, i.e. the physical structure and the functional organisation of the digital city (Dyer et al., 2017). A contribution from Schaffers et al. (2012) identified four main phases in this transformation, since the launch of *World Wide Web Consortium* in 1994. In the first phase, some traditional city functions were reinforced and amplified by their transfer to a web environment and have been opened to non-experts *via* the creation of static web pages that provided information about the urban area with text, maps and pictures. Representation and visualisation (e.g. 2D or 3D graphic interfaces) technologies characterised the architecture of this first augmentation of the city space. About ten years later, a second phase was made possible by the diffusion of the ADSL communication bandwidth, the development of open-source Content Management Systems and web publishing platform built on PHP language and the MySQL database (Schaffers et al., 2012). These have been fundamental ingredients for the birth of the interactive web (i.e. the Web 2.0) in which users are both creators and consumers of web content and applications. Interestingly, the web2.0 materialised in the virtual dimension a fundamental trait of the "city as social space of agglomeration and collaboration" (Schaffers et al., 2012, p. 125). Thereafter, the virtual space of the city has rapidly become crowded with collaborative platforms (e.g. *Scoop.it*, *SeeClickFix*), wikis, blogs, social networking sites, media sharing, hosting of web applications, mash-ups and similar. This double nature, physical and virtual, of the city becomes evident with the living labs that build upon the belief that collective intelligence and social capital, organised through crowdsourcing processes (Certomà & Rizzi, 2017), are key for problem-solving and place-based innovation (Gloor et al., 2008). At the end of the 2010s, a third phase saw the emergence of embedded systems, a combination of computer hardware and software, either designed or at least programmable for a specific function (Kumar & Nath, 2019) and wireless networks. These made the city's metabolism parameters measurable thanks to the collection of data realised through diffuse sensors and interconnected smart devices, and the subsequent elaboration of results (Du et al., 2019). The city itself became the laboratory for experimenting with the IoT and a provider of services for citizens created by citizens through the use of their mobile devices and the availability of augmented reality applications (Mossberger et al., 2013; Tselentis et al., 2010). Today, a further evolution of this unprecedented availability of urban Big Data is the request for their openness, which "demands open innovation models and people-driven innovation models to turn capabilities offered by data and technologies to services and solutions" (Schaffers et al., 2012, p. 24). The adoption of digital technologies for addressing social problems constituted a normalisation of practices that were regarded as innovative and eccentric up to recent time. It's most pervasive, still commonly unnoticed consequence, is the emergence of digital governmentality (Dean, 1999; Rajagopal, 2014), in which whose predictive analytics are used as new technologies for the measurement of population dynamics, and whose constant incitation to action replaced traditional mechanisms for impulses and desires control adopted in traditional governmentality processes of modern States (Barry, 2019) (see Chapter 4).

Therefore, the digital revolution restructured power geometries between cities and among citizens (de Wall, 2015), characterised by a major fracture between those that are in a position to control and modify the codes that govern our social (and private) life, those who passively use them, and those who have no access to digital devices and infrastructure at all (see Chapter 5). Notably, the resulting digital divide (Van Dijk & Hacker, 2003) is not limited to the access of technological infrastructure and devices but is also connected with cultural and social barriers such as digital literacy, education or language issues (Norris, 2003; Selwyn, 2004; Warschauer, 2004).

Together with producing dramatic transformations in the production of space and urban life practices, the digital revolution also generated different spatialities by multiplying the possibilities of people's encounters with space because the digital dimension hosts "new spatial relations on multiple layers" (Gairola & Roth, 2019). The concept of spatiality refers to a relative and relational idea of space produced by social interrelations and interactions with it (Sheppard, 2002). Different to the idea of space as a physical system of coordinates (adopted, for instance, in the research referred to in Sect. 2.4), the idea of spatiality opens the way to consider different modalities through which space takes effect on political, economic and cultural systems (Massey, 2005). Digital spatialities spring from the encounter of society with space mediated by digital tools and processes, which is, increasingly, our daily experience of space (Ash, 2009; Sutko et al., 2010). These new spatialities have been differently named as code/spaces, hybrid spaces (de Souza e Silva, 2006), digiplace (Zook & Graham, 2007), net locality (Gordon et al., 2011), augmented reality (Graham et al., 2013), mediated spatiality (Leszczynski, 2016)—to mention but a few. They refer to "the instantiation of digital networks as internet exchanges, data centres, fibre optic cables and their landing sites, as well as [to] the contentious economic, social, political, and historical contexts of their geographies" (Ash et al., 2018, p. 33); as well as to the "spatialities of algorithms themselves, i.e. the geographies of their coding, circulation, and appropriation" (Ash, 2018, p. 34).

3.6 A Critical Research Agenda on Digital Social Innovation

It was the polyvocal vocation of critical geography (Blomley, 2006) that inspired the research lines of this book. These do not add up to a single picture but pinpoint aspects of DSI in need of more careful consideration, suggested from critical scholarly research. Investigation on socio-spatial representations are based on post-modern critical geography contributions (Peet, 1977) that revolve around issues of symbolic and material power and its spatial configurations (Crampton & Elden, 2007).

Poststructuralist analyses on the social construction and reception of technological innovations offers a standpoint for analysing social (re)production practices in urban contexts, particularly with regard to the generation of inequalities and domination structures (Hubbard et al., 2002) through governance processes (see Chapter 4). Here, critical digital studies (Kroker & Kroker, 2013; Lanier, 2006) contribute to pondering how digital technologies influence the practices of society and space production, its contestation and transformation against the backdrop of digital global geopolitics. The suggestions lead us to investigate the socio-political underpinning of governance processes in the digital age (de Rosnay, 2006; Hunsingeret al., 2019; McFarlane & Söderström, 2017) (see Chapter 5); and make it imperative to consider under what conditions DSI initiatives can become emancipatory political gestures and produce actually innovative forms of collective agency against the neoliberal hyperconnected-city. These last conditions require transitioning from controlled to open data-driven models; from a business-led and techno-deterministic condition to socially innovative community-driven cities; and from technological devices designed on the basis of corporate interest to those based on people's needs (Calzada & Cobo, 2015).

The research lines explored in the following chapters hence aggregate into the following three clusters.

- *Representations: how the symbolic production of DSI entwines with urban narratives.*

Representations include narratives, visions and discourses that attribute meaning and values to collective imaginaries; and often make thinkable ideas, practices and events that have not been thought of (or were even unthinkable) before. Not surprisingly, most of the innovative capacity of DSI communities primarily feed on the ability to generate novel and powerful collective understandings because "rather than [practically] invent a new type of city, the extraordinary array of smart technologies available allow existing spaces to be reconfigured, experienced and imagined in new ways" (Han & Hawken, 2018, p. 2). DSI processes are enacted through ad-hoc narratives (e.g. collective intelligence or enabling-technology; Turner, 2006), imaginaries (e.g. the "punk-internet activists' approach" (Harris, 2018) or the *EU Next Generation Internet* plan), and visions (e.g. the smart city, the people friendly city, the resilient city paradigms and similar). The strength of DSI representations root in two assumptions which are among the key tenets that made the digital turn possible. These include the aspiration for greater direct participation in democratic processes and the ideal of a collaborative society of expert citizens operating through decentralised and connected platforms. Moreover, different social imaginaries hold a connection with various technological tools; and contribute to the constitution of distinctive urban contexts supporting or impeding changes in society and space. The creation of these different contexts is, in turn, fuelled by diverse visions of the city that create new forms of spatial attachments, vehiculated or actively contributed via digital technologies.

- *Reproduction: how DSI practices reproduce themselves and the city.*

While a prolific research line in critical geography attempted to chart digital space, to examine the (bio)politics of algorithms (Dodge and Kitchin, 2008; Graham et al., 2013; Thrift & French, 2002), and to criticise the smart city paradigm (Greenfield, 2017; Verrest & Pfeffer, 2019), the (re)production processes of the augmented city and the role of DSI processes have only been partially explored. DSI initiatives, emerging at the crossroads between the rapid evolution of digital innovations and the cogency of social challenges, clearly show how the transformations of the coded set of procedures that make a city work can be impacted by different actors. This calls for the consideration on how the social production of space is mediated through the social construction of technologies (Glimell, 2001) and through the reproductive processes of places where DSI happen. At the same time, in DSI processes digital technologies work as social technologies, which induce specific collective behaviours through the combination of power (the practices, mechanisms and technologies that constitute authority) and knowledge (the forms of thought and expertise used to frame and inform the process of governing).

- *Power: the socio-political implications of DSI practices.*

Existing DSI initiatives are very diverse, rooted in opposing world-views and aspire to the realisation of different ideals, directly connected with the political value of adopted technologies. The mainstream approach to DSI (Sect. 2.3) considers the promoted technical and organisational innovations as functional to facilitating the automation of tasks which improve the quality and efficiency of business or government processes (Misuraca et al., 2017). A radically different approach informs DSI initiatives of digital activists. These are endowed with the intention of subverting the existing structure of digitally mediated governance (see Chapter 5). Such a revolutionary understanding is regarded by many as the very nature of DSI, with the aim to change the socio-technological cognitive frames of reference and alter current social systems by working outside institutional settings. Walking on the edge between functionalist or subversive practices, many European digital social innovators enriched the debate by proposing a reformist perspective in line with European core values of public accountability, social justice and democratisation of innovation. While rarely attempted, it is essential to distinguish between these different forms of DSI because it allows us to appreciate how these processes play opposite roles in the reproduction or contestation of digital capitalism and the subsequent discursive, political and material transformation of urban spaces.

As a matter of fact, innovation for society can, in particular circumstances, become "socially divisive; have unintended consequences that have negative social effects (by excluding people who are affected by the innovation in the design and implementation stages); and become vulnerable to co-option and/or mission drift" (Caulier-Grice et al., 2012, p. 46). These considerations also apply to DSI. A thorough understanding of the scarcely investigated geopolitics of DSI or of the multiple manifestations of power geometries in DSI processes cannot be limited to issues of wide public access to technologies and networks but requires reflection on how deep-rooted patterns of social inequalities are reinforced or transformed in the digital age.

A systematic social critique of the DSI phenomenon has not been elaborated upon until now, particularly with the relevant significance placed upon the urban context in which it manifests. Considering the fast diffusion of DSI initiatives, their versatility and pervasiveness in virtually all spheres of social life, it is time to engage with critical consideration of their nature, characteristics and implications.

To this end, critical geographers' elaborations on the impact of the digital revolution on urban space and society can work as guiding lines for an exploration of the reproduction, representation and power processes connected with DSI. The deconstruction of social representations in and through DSI can provide us with new insights on their meanings, roles and objectives. Behind the imaginary of the hyperconnected-city, the existence of multiple, diverse representations of DSI processes permeate the urban fabric. These reveal that, further than working as a laboratory for experimenting technocratic solutions lead by digital capitalism, the city is also an incubator for citizens' critical engagement, which can detect and defuse the unwanted consequences of the digital revolution in general and of neoliberal forms of DSI. An exploration of the different approaches to DSI and the projects these advance in the (re)productive process of urban space and society, signal the presence of diverse intentions. These are supported by alternative representations, performed through alternative governance processes and fuelled by opposite flows of power. The goal of re-politicising the hegemonic logic of the digital turn (Agyemanm, 2015; Cardullo & Kitchin, 2018) requires attentive consideration of the conditions under which DSI initiatives can be emancipatory political gestures. Therefore, the supposedly progressive, democratic and empowering character of DSI can be problematised in light of the techno-material practices adopted, their accessibility, and their socio-political impacts.

Along this line of thought, the following chapters propose an analysis of how DSI processes are produced and circulate in society; how they enhance and maintain specific spatial configurations; how technological practices associated with DSI initiatives chart discourses, infiltrate material methods in a relational, contingent and contextual way and how they sustain particular kinds of interests in society, by feeding existing geometries of power or creating new ones. Eventually, we question whether DSI is actually adequate to fuel the transition from a business-led, techno-deterministic city towards a socially innovative community-driven one—instead of being designed on the basis of corporate interests, can it be created for people's needs?

References

Agyemanm, J. (2015). *Sharing cities: A case for truly smart and sustainable cities*. MIT Press.

Alvord, S. H., Brown, L. D., & Letts, C. W. (2004). Social entrepreneurship and societal transformation: An exploratory study. *The Journal of Applied Behavioral Science, 40*(3), 260–282.

Ash, J. (2009). Emerging spatialities of the screen: Video games and the reconfiguration of spatial awareness. *Environment and Planning A, 41*(9), 2105–2124.

Ash, J., Kitchin, R., & Leszczynski, A. (2018). Digital turn, digital geographies? *Progress in Human Geography, 42*(1), 25–43.

Baker, T., Baeck, P., & Stokes, M. (2017). What's next for digital social innovation? A deliverable of H2020 project DSI4EU, Nesta.

Barcelona Activa. (2018). *Mapping DSI—Cities and urban development*. Available at https://digitalsocial.eu/images/upload/33-Cities%20and%20urban%20development.pdf.

Barroso, M. (2014). *Political guidelines*. Available at http://ec.europa.eu/commission_2010-2014/president/pdf/press_20090903_en.pdf.

Barry, L. (2019). The rationality of the digital governmentality. *Journal for Cultural Research, 23*(4), 365–380.
Bason, C. (2010). *Leading public sector innovation: Co-creating for a better society*. University Press.
Bates, S. (2012). *The social innovation imperative. Create winning products, services, and programs that solve society's most pressing challenges*. The McGraw-Hill Companies.
Blomley, N. (2006). Uncritical critical geography? *Progress in Human Geography, 30*(1), 87–94.
Boelman, V., & Heales, C. (2015). Social innovation strategies—Regional report. SI-DRIVE, European Commission. Available at http://www.si-drive.eu/wp-content/uploads/2018/03/D3.6-SI-DRIVE-Global-Region-Report-2015.pdf.
Bommert, B. (2010). Collaborative innovation in the public sector. *International Public Management Review, 11*, 15–33.
Brandsen, A., Cattacin, T., Evers, S., & Zimmer, A. (Eds.). (2016). *Social innovations in the urban context*. Springer.
Brenner, N. (2009). What is critical urban theory? *City, 13*(2–3): 198–207.
Bria, F. (2014). Digital Social Innovation. DSI4EU project Interim report, Nesta—EU H2020. Available at https://waag.org/sites/waag/files/media/publicaties/dsi-report-complete-lr.pdf.
Bria, F. (2015). *Growing a Digital Social Innovation ecosystem for Europe* (Digital Social Innovation Final Report). Available at https://www.nesta.org.uk/sites/default/files/dsireport.pdf
Bučar, M., & Rissola, G. (2018). *Place-based innovation ecosystems: Ljubljana start-up ecosystem and the technology park Ljubljana*. European Commission. Available at https://op.europa.eu/en/publication-detail/-/publication/3e0ebc68-358b-11e9-8d04-01aa75ed71a1/language-en.
Cairncross, F. (1997). *The death of distance*. Orion Business Books.
Cajaiba-Santana, G. (2014) Social innovation: Moving the field forward. A conceptual framework. *Technological Forecasting and Social Change, 82*, 42–51.
Calzada, I., & Cobo, C. (2015). Unplugging: Deconstructing the smart city. *Journal of Urban Technology, 22*(1), 23–43.
Cardullo, P., & Kitchin, R. (2018). Smart urbanism and smart citizenship: The neoliberal logic of 'citizen-focused' smart cities in Europe. *Environment and Planning C: Politics and Space, 37*(5), 813–830.
Castells, M. (1996). *The rise of the network society*. Blackwell.
Caulier-Grice, J., Davis, A., Patrick, R., & Norman, W. (2012). *Defining Social Innovation*. TEPSIE, European Commission Brussels.
Cels, S., de Jong, J., & Nauta, F. (2012). *Agents of change: Strategy and tactics for social innovation*. Brookings Institute Press.

Certomà, C., & Rizzi, F. (2017). Crowdsourcing process for citizen-driven governance. In C. Certomà, M. Dyer, F. Rizzi, & L. Pocatilu (Eds.), *Citizen empowerment and innovation in the data-rich city* (pp. 57–77). Springer.

Crampton, J. W., & Elden, S. (Eds.). (2007). *Space, knowledge and power: Foucault and geography*. Ashgate.

Dawson, P., & Daniel, L. (2010). Understanding social innovation: A provisional framework. *International Journal of Technology Management, 51*(1), 9–12.

De Cindio, F., & Aurigi, A. (2008). *The augmented urban space*. Routledge.

de Rosnay, M. D. (2006). Alternative policies for alternative Internets. *Journal of Peer Production, 9*, 1–10.

de Souza e Silva, A. (2006) From cyber to hybrid: Mobile technologies as interfaces of hybrid spaces. *Space and Culture, 9*(3), 261–278.

de Waal, M. (2014). The city as interface: How new media are changing the city. Rotterdam: NAi010 Publishers.

Dean, M. (1999). *Governmentality: Power and rule in modern society*. Sage.

Dodge, M., & Kitchin, R. (2008). *Atlas of cyberspace*. Available at http://www.kitchin.org/atlas/.

Drewe, P., Klein, J., & Hulsbergen, E. (Eds.). (2008). *The challenge of social innovation in urban revitalization*. Techne Press.

Drydyk, J. (2013). Empowerment, agency, and power. *Journal of Global Ethics, 9*(3), 249–262.

Du, R., Santi, P., Xiao, M., Vasilakos, A., & Fischione, C. (2019). The sensable city: A survey on the deployment and management for smart city monitoring. *IEEE Communications Surveys & Tutorials., 21*(2), 1533–1560.

Dyer, M., Gleeson, D., & Grey, T. (2017). Framework for collaborative urbanism. In C. Certomà, M. Dyer, L. Pocatilu, & F. Rizzi (Eds.), *Citizen empowerment and innovation in the data-rich city*. Springer tracts in civil engineering. Springer.

EC. (2005). A European information society for growth and employment. Communication from the Commission to the Council, the European Parliament, The European Economic and Social Committee and the Committee of the Regions. Brussels 1.6.2005. COM(2005) 229 final.

EC. (2008). *'Renewed Social Agenda' adopted by the commission in July 2008*. Available at http://ec.europa.eu/social/main.jsp?catId=89&langId=en&newsId=984&furtherNews=yes.

EC. (2010a). *Lisbon strategy evaluation document*. Available at http://ec.europa.eu/transparency/regdoc/index.cfm?fuseaction=list&coteId=2&year=2010&number=114&language=EN.

EC. (2010b). Europe 2020 Flagship Initiative Innovation Union SEC (2010) 1161. Available at http://ec.europa.eu/research/innovation-union/pdf/innovation-union-communication_en.pdf.

EC. (2011). *White Paper on European Governance, 2001.* http://ec.europa.eu/governance/index_en.htm.

EC. (2013). *Expert workshop report: Research and innovation at EU level under Horizon 2020 in support of ICT-driven public sector innovation, Brussels on 31 January 2013.* Available at http://ec.europa.eu/information_society/newsroom/cf/dae/document.cfm?doc_id=2589market/en/news/future-internet-%E2%80%93-public-private-partnership-fi-ppp-0.

EC. (2014, November). Digital agenda for Europe Luxembourg: Publications Office of the European Union.

EC. (2020). *Next generation internet 2020.* European Commission. https://www.ngi.eu/about/.

Eckhardt, J., Kaletka, C., & Pelka, B. (2016). Copy here, paste there? On the challenges of scaling inclusive social innovations. In M. Antona & C. Stephanidis (Eds.), *Universal access in human-computer interaction: Methods, techniques, and best practices.* Springer

EU Science Hub. (2020). *Smart specialisation.* European Commission. Available at https://ec.europa.eu/jrc/en/research-topic/smart-specialisation.

Evans, J., & Karvonen, A., et al. (2010). Living laboratories for sustainability: Exploring the politics and epistemology of urban transition. In H. Bulkeley (Ed.), *Cities and low carbon transitions* (pp. 126–141). Routledge.

Friedman, T. (2005). *The world is flat: A brief history of the twenty-first century.* Straus and Giroux.

Gairola, R. K., & Roth, M. (2019). Cyber zones: Digital spatialities and material realities across Asia. *Asiascape: Digital Asia, 6*(1–2), 4–16.

Glimell, H. (2001). *The social production of technology: On the everyday life with things.* BAS Publisher.

Gloor, P. A., Paasivaara, M., Schoder D., & Willems P. (2008). Finding collaborative innovation networks through correlating performance with social network structure. International Journal of Production Research: Vol. 46, Enhancing Performance in Industrial Collaborative Networks-a Selection of Papers from the PRO VE'06 Conference, pp. 1357–1371.

Goodchild, M. F. (2007). Citizens as sensors: The world of volunteered geography. *GeoJournal, 69*, 211–221.

Gordon, E., & De Souza e Silva, A. (2011). *Net locality: Why location matters in a networked world.* Wiley-Blackwell.

Graham, M., Zook, M., & Boulton, A. (2013). Augmented reality in the urban environment: Contested content and the duplicity of code. *Transactions of the Institute of British Geographers., 38*(3), 464–479.

Granstrand, O., & Holgersson M. (2020). Innovation ecosystems: A conceptual review and a new definition, Technovation 90/91.

Greenfield, A. (2017). *Radical technologies: The design of everyday life.* Verso.

Hearn, G., Foth, M., & Stevenson T. (Eds.). (2011). Community engagement for sustainable urban futures. *Futures, 43*(4), 357–360.

Hall, P., & Pfeiffer, U. (2013). *Urban future 21.* E&FN Spon.

Hämäläinen, T., & Heiskala, R. (2007). *Social innovations, institutional change and economic performance (edited in association with SITRA, the Finnish innovation fund).* Edward Elgar.

Han, H., & Hawken, S. (2018). Introduction: Innovation and identity in next-generation smart cities. *City, Culture and Society, 12*, 1–4.

Harris, J. (2018). The punk-rock Internet. *The Guardian*, February 1. Available at https://www.theguardian.com/technology/2018/feb/01/punk-rock-Internetdiy-rebels-working-replace-tech-giants-snooperscharter.

Harvey, D. (1989). *The condition of postmodernity.* Sage.

Heynen, N. (2014). The urban century. *Progress in Human Geography, 38*(4), 598–604.

Hubbard, P., Kitchin, R., Bartley, B., & Fuller, D. (2002). *Thinking geographically: Space, theory and contemporary human geography.* Continuum Books.

Hunsinger, J., Allen, M. M., & Klastrup, L. (Eds.). (2019). *Second international handbook of Internet research.* Springer.

Ibrahim, S., & Alkire, S. (2007). Agency and empowerment: A proposal for internationally comparable indicators. *Oxford Development Studies, 35*(4), 379–403.

INEA. (2020). *Smart cities & communities.* Available at https://ec.europa.eu/inea/en/horizon-2020/smart-cities-communities.

JPI. (2015). *Urban Europe, 2015, JPI Urban Europe launches SRIA 2.0 and strengthens its relevance to urban policy agendas.* https://jpi-urbaneurope.eu/jpi-urban-europe-launches-sria-2-0-and-strengthens-its-relevance-to-urban-policy-agendas/.

JRC. (2020). *Cities as innovation hubs.* Available at https://urban.jrc.ec.europa.eu/thefutureofcities/cities-as-innovation-hubs#the-chapter.

Komninos, N. (2009). Intelligent cities: Towards interactive and global innovation environments. *International Journal of Innovation and Regional Development, 1*(4), 337–355.

Kroker, A., & Kroker, M. (2013). *Critical digital studies: A reader.* UTP.

Kumar, A., & Nath, V. (2019). Study and design of smart embedded system for smart city using Internet of things. In V. Nath & J. Mandal (Eds.), *Nanoelectronics, circuits and communication systems* (Vol. 511). Lecture Notes in Electrical Engineering. Springer.

Landi, D. (2019). The image of the hyper city. *International Journal of Semiotic Law, 32*, 533–548.

Landry, C. (2015). Cities as ambition, Comedia.

Lanier, J. (2006). Digital maoism: The hazards of the new online collectivism. *Edge magazine*. Available at https://www.edge.org/conversation/jaron_lanier-digital-maoism-the-hazards-of-the-new-online-collectivism.

Leszczynski, A. (2016). Speculative futures: Cities, data, and governance beyond smart urbanism. *Environment and Planning A*. Epub ahead of print 22 May 2016.

MacCallum, D., Moulaert, F., Hillier, J., & Vicari, S. (Eds.). (2009). *Social innovation and territorial development*. Ashgate.

Maglavera, T., Niavis, H., Moutsinas, G., Passani, A., & De Rosa, S. (2019). *Digital transformation for a better society*. ChiC, European Commission. Available at https://capssi.eu/wp-content/uploads/ChiC_D5.2_Digital_Transformation_for_a_better_society-whitepaper.pdf.

Mair, J., & Ganly, K. (2010). Social entrepreneurs: Innovating towards sustainability. In: E. Assadourian (Ed.), *State of the World 2010, transforming cultures: From consumerism to sustainability*. Worldwatch Institute.

March, H., & Ribera-Fumaz, R. (2014) Smart contradictions: The politics of making Barcelona a self-sufficient city. *European Urban and Regional Studies, 23*(4), 816–830.

Marcuse, P., Imbroscio, D., Parker, S., & Davier, J. (2014). Critical urban theory versus critical urban studies: A review debate. *International Journal of Urban and Regional Research, 38*(5), 1904–1917.

Massey, D. (2005). *For space*. Sage.

Massey, J., & Snyder, B. (2015). The hypercity that occupy built. In J. Geiger (Ed.), *Entr'acte. Avant-gardes in performance*. Palgrave Macmillan.

McFarlane, C., & Söderström, O. (2017). On alternative smart cities: From a technology-intensive to a knowledge-intensive smart urbanism. *City, 21*(3–4), 312–328.

Millard, J., & Carpenter, G. (2014). Digital technology in social innovation, A deliverable of the project: The theoretical, empirical and policy foundations for building social innovation in Europe (TEPSIE), European Commission- 7th Framework Programme. European Commission, DG Research.

Misuraca, G., Pasi, G., & Urzi Brancati, C. (2017) *ICT-enabled social innovation: Evidence & prospective*. Publications Office of the European Union.

Mossberger, K., Tolbert, C., & Franko, W. (2013). *Digital cities: The Internet and the geography of opportunity*. Oxford University Press.

Moulaert, F., Martinelli, F., Swyngedouw, E., & Gonzalez, S. (2005). Towards alternative model(s) of local innovation. *Urban Studies, 42*, 1969–1990.

Moulaert, F., & Van Dyck, B. (2013). Framing social innovation research: A sociology of knowledge perspective. In F. Moulaert, D. MacCallum, A. Mehmood & A. Hamdouch (Eds.), *International handbook on social innovation*. Edward Elgar.

Mulgan, G. (2006). The process of social innovation. Innovations: Technology, governance, globalization, *1*(2), 145–162.

Murdoch, J. (2005). *Post-structuralist Geography*. London: SAGE.

Murray, G., Caulier-Grice, J., & Mulgan, R. (2010) *Open book of social innovation*. The Yong Foudation/Nesta.

Nielsen J. (2006, October 9). The 90–9–1 rule for participation inequality. Social media and online communities. Nielsen Norman Group available at https://www.nngroup.com/articles/participation-inequality/.

Nicholls, A., Simon, J., & Gabriel, M. (2015) Introduction: Dimensions of social innovation. In *New frontiers in social innovation research* (pp. 1–26). Palgrave Macmillan.

Norris, P. (2003). Preaching to the converted?: Pluralism, participation and party websites. *Party Politics, 9*(1), 21–45.

Oosterlynck, S., Kazepov, Y., Novy, A., Cools, P., Barberis, E., Wukovitsch, F., Sarius, T., & Leubolt, B. (2013). *The butterfly and the elephant: local social innovation, the welfare state and new poverty dynamics [WWW document]* (ImPRovE Discussion Paper 13/03). http://improve-research.eu. Accessed 30 May 2018.

Ozman, M., & Gossart, C. (2019). Digital Social Innovation: Exploring an emerging field. ISIRC 2018 Conference, 5–7 September, Heidelberg. Available at SSRN: https://ssrn.com/abstract=3434363 or http://dx.doi.org/https://doi.org/10.2139/ssrn.3434363.

Peet, R. (1977). *Radical geography: Alternative viewpoints on contemporary social issues*. Chicago: Maaroufa Press.

Rajagopal, I. (2014). Does the Internet shape a disciplinary society? The information-knowledge paradox. *First Monday, 19*(3).

Ratti, C., & Townsend, A. (2011, September). The social nexus. *Scientific American*, 30–35.

Rissola, G., Hervas, F., Slavcheva, M., & Jonkers, K. (2017). *Place-based innovation ecosystems: Espoo Innovation Garden and Aalto University (Finland)*. European Union. Available at https://op.europa.eu/en/publication-detail/-/publication/141a3f98-1e79-11e7-aeb3-01aa75ed71a1/language-en.

Salamon, L. M., Sokolowski S. W., & List, R. (2003). *Global civil society: An overview*. Center for Civil Society Studies Institute for Policy Studies, the Johns Hopkins University.

Schaffers, H., Ratti, C., & Komninos, N. (2012). Special issue on smart applications for smart cities—New approaches to innovation: Guest editors' introduction. *Journal of Theoretical and Applied Electronic Commerce Research, 7*(3).

Selwyn, N. (2004). Reconsidering political and popular understandings of the digital divide. *New Media & Society, 6*(3), 341–362.

Seyfang, G., & Smith, A. (2007). Grassroots innovations for sustainable development: Towards a new research and policy agenda. *Environmental Politics, 16*, 584–603.

Sheppard, E. (2002). The spaces and times of globalisation: Place, scale, networks and positionality. *Economic Geography, 78*(3), 307–330.

SI-DRIVE. (2017). *Towards a general theory and typology of social innovation*. SI-DRIVE, European Commission.

Simon, J., Bass, T., Boelman, V., & Mulgan, G. (2017). *Digital democracy the tools transforming political engagement*. NESTA. Available at https://www.nesta.org.uk/sites/default/files/digital_democracy.pdf.

Soja, E. (1989). *Postmodern geographies: The social reassertion of space in critical theory*. London: Verso Press.

Sörensen, E., & Torfing, J. (2011). Enhancing collaborative innovation in the public sector. *Administration & Society, 43*, 842–868.

Stern, N. (2007). *The Economics of climate change: The stern review*. London: HM Treasury. Available at https://webarchive.nation-alarchives.gov.uk/+/http://www.hm-treasury.gov.uk/stern_review_climate_change.htm.

Stokes, M. (2020). *What system factors help DSI to grow and thrive?* DSI4EU, Nesta. Available at https://digitalsocial.eu/blog/145/a-first-run-of-the-dsi-index.

Sutko, D., De Souza, M., & Silva, A. (2010). Location-aware mobile media and urban sociability. *New Media & Society, 13*, 807–823.

Thompson, M. (2019). Playing with the rules of the game: Social innovation for urban transformation. *International Journal of Urban and Regional Research, 43*(6).

Thrift, N., & French, S. (2002). The automatic production of space. *Transactions of the IBG, 27*(3), 309–335.

Tselentis, G., Galis, A., Gavras, A., Krco, S., Lotz, V., Simperl, E., Stiller, B., & Zahariadis, T. (Eds.). (2010). *Towards the future Internet—Emerging trends from European research*. IOS Press.

Van der Have, R. P., & Rubalcaba, L. (2016). Social innovation research: An emerging area of innovation studies? *Research Policy, 45*(9), 1923–1935.

van Dijk, J. A. G. M., & Hacker, K. (2003). The digital divide as a complex and dynamic phenomenon. *Information Society, 19*(4), 315–326.

Valentina, A. L., Maria, A. R., Carmelo, A. T., Davide, A. U., Ricardo, R. B., Filipe, B. E., Peter, B. E., Paolo, B. E., Flavio, B. O., Ioris, B. U., & Sandra, L.C. (2019). *The future of cities—Opportunities, challenges and the way forward*. European Commission. Available at https://ec.europa.eu/jrc/en/facts4eufuture/future-of-cities.

Vanolo, A. (2013). Smartmentality: The smart city as disciplinary strategy. *Urban Studies, 51*(5), 883–898.

Verona, G., Prandelli, E., & Sawhney, M. (2006). Innovation and virtual environments: Towards virtual knowledge brokers. *Organization Studies, 27*(6), 765–788.
Verrest, H., & Pfeffer, K. (2019). Elaborating the urbanism in smart urbanism: Distilling relevant dimensions for a comprehensive analysis of Smart City approaches. *Information, Communication and Society, 22*(9), 1328–1342.
Warschauer, M. (2004). *Rethinking the digital divide*. MIT Press.
Westley, F., McGowan, K., & Tjornbo, O. (2017). *The evolution of social innovation: Building resilience through transitions*. Edward Elgar.
Whittle, J., Ochu, E., Ferrario, M. A., Southern, J., & McNally, R. (2012). *Beyond research in the wild: Citizen-led research as a model for innovation in the digital economy*. Digital Futures 2012 Conference. Available at https://www.academia.edu/3297686/Beyond_research_in_the_wild_Citizen-Led_research_as_a_model_for_innovation_in_the_Digital_Economy.
Zook, M. A., & Graham, M. (2007). Mapping DigiPlace: Geocoded internet data and the representation of place. *Environment and Planning B: Planning and Design, 34*(3), 466–482.

Links

Citizen Foundation. https://citizens.is/.
D-Cent Decentralised Citizens ENgagement Technologies. https://dcentproject.eu/.
Digital Social Innovation for Europe—DSI4EU. https://cordis.europa.eu/project/id/688192.
Nappy Napuri. https://holvi.com/shop/nappinaapuri/.
Open Source software society. https://gfoss.eu/.
SISCODE. https://siscodeproject.eu/.
Social Innovation: Driving Forces of Social Changes (SI-DRIVE). https://www.si-drive.eu
TEPSIE. https://cordis.europa.eu/project/id/290771/reporting.
WeProductise. https://weproductise.com/.

CHAPTER 4

Representation: The Social Imaginaries of Digital Social Innovation

Abstract This chapter analyses the diverse and opposite social imaginaries associated with the collective representation and the practical realisation of Digital Social Innovation (DSI) initiatives. It considers representations emerging from discourses, narratives or visions that substantiate collective imaginary as the initial moment in the generation of new social configurations. We can thus appreciate how different imaginaries of DSI, associated with urban imaginaries, increased in recent decades when the digital turn radically changed urban functional processes and societal behaviours. The chapter explains how the meaning and characteristics of DSI are not stabilised yet, and alternative competing definitions are adopted in different socio-political and economic macro-regional contexts and communities of practice. From the approach functional to the reproduction of the current socio-economic order to the more reformist and even the revolutionary ones, different understandings of DSI hold a tight connection with the imaginaries of contemporary urbanity (here described as the hyperconnected, receptive and do-it-yourself city).

Keywords Digital social innovation · Socio-technical imaginaries · Urban narratives · CivicTech · Functionalist · Reformist · Revolutionary innovation · Hyperconnected · Receptive and do-it-yourself city

C. Certomà, *Digital Social Innovation*,
https://doi.org/10.1007/978-3-030-80451-0_4

4.1 Introduction

In the previous chapters of this book, DSI has been presented as a complex and multifaceted phenomenon enabling new socio-technological processes that are able to impact on the organisation and functioning processes of urban assemblages. Most of the innovative potential of DSI initiatives are embedded in their capacity to offer us a new gaze on reality. This can eventually lead us to discover unexplored routes to new initiatives for addressing collective problems in more efficient, effective, democratic or sustainable ways than before. These initiatives can be easily regarded as the materialisation of specific techno-scientific rationalities which serve diverse political visions and are conveyed *via* social imaginaries and popularised through ad-hoc narratives.

For quite some time, the ephemeral nature of digital agency has prevented the attempts at analytically exploring its associated practices and encapsulating them into stable definitions, with generalisable functional attributes and identifiable routines. Despite the volatile and fast-evolving character of web-based innovation, a general understanding of interactive digital technologies relies on identifiable worldviews and consolidated socio-economic paradigms. To contribute towards unveiling them, this chapter shows how DSI initiatives are differently named in different cultural, economic and geographical contexts and also that these different definitions underlay different worldviews and diverse readings of what is the bad and good for society as a whole by prefigurating alternative social ordering and ends.

Being a social technology (Chapter 1), DSI (and its homologues) aim towards the generation of new social configurations, endowed with distinguishing political and cultural visions. With the aim to better disentangle the interpretative flexibility that characterises the DSI domain (see Chapter 1), this chapter analyses the diverse and opposite social imaginaries associated with the collective representation and the practical realisation of DSI initiatives.

By adopting a deconstructionist approach, we consider here representations expressed by discourses, narratives or visions that substantiate collective imaginary as the very origin of social processes. By placing these diverse discursive representations in their contexts of realisation, and applying the perspective of critical geography (as described in Chapter 2), we are able to more appropriately account for the intertwining of material and symbolic spaces of agency. This allows us to appreciate how different imaginaries of DSI, associated with urban imaginaries, proliferated in recent decades when the digital turn radically changed urban functional processes and societal behaviours. To apply this interpretative framework requires a journey into considering what are the urban functional processes and how are imaginaries enacted through specific processes of governance.

4.2 The Social-Technical Imaginaries

Working upon the seminal research of cultural imaginaries theorists, Sheila Jasanoff defined socio-technical imaginaries as "collectively held, institutionally stabilized, and publicly performed visions of desirable futures, animated by shared understandings of forms of social life and social order attainable through, and supportive of, advances in science and technology" (Jasanoff, 2015b, p. 4). Similar to discourses, imaginaries are collective and systemic, but unlike discourses they also materialise into actions or artefacts—notably technological ones, further than on linguistic expressions (Jasanoff, 2015a, p. 20). Unlike mere ideas, their performative character resides in the capacity to turn speculative plans into practical realisations. Imaginaries operate in the interstitial space between discourses and decision, imagination and action and between public opinion and state policy (Jasanoff & Kim, 2015). They are able to make plans real. Imaginaries can work as transformative forces because they are, by their very nature, intended to be embedded and incorporated into the design of policies and material systems.

The performative process that turns imaginaries into factual reality is fuelled by the narratives which allow ideas to "gain traction, acquire strength and cross scales" in society at large (Sovacool & Hess, 2017, p. 17). Narratives are stories that help people to make sense of their experiences and inspire specific understandings of reality to foresee future developments (Wittmayer et al., 2019). Social imaginaries are the object around which a narrative elaborates and restitutes people with a logical or emotional ordering structure for their knowledge and expectations and a justification for their agency. In fact, imaginaries are also filtered by social values, and thus bear a relation with morality because they mobilise collective utopian or dystopian prefiguration of possible future social transformations and the way towards them. Unavoidably, in a lively society, multiple and diverging imaginaries can both coexist in a dialectic relationship or even clash. Individual preference means that one person's utopia might be someone else's dystopia (Sovacool & Ramana, 2015). And, imaginaries are even more powerful because they do not only encode visions of attainable futures but also provide indications on what life ought to be, in reference to a society's shared understanding of good and evil (Jasanoff, 2015b). To do so, imaginaries "articulate how normative commitments to social life and order are reflected in the construction of techno-scientific pursuits at the multiple levels of design, system implementation, and policy development in pursuit of the good life" (Tidwell & Tidwell, 2018, p. 103). They are pervasive in policies, communication, legal and normative productions, art products or even religious prescriptions, and, consequently, they cannot but be recognised as crucial in the exercise of power.

Imaginaries of SI have been key components in societal transformation since the XIX century when it was pursued through revolutionary movements and was therefore closely associated with prefigurative ideals of social subversion. Later on, the innovation element gained a central role with greater power attributed to the capacity of novel ideas to driving social changes (Segercrantz & Seeck, 2013). In this context, the importance of re-meaning the past, of providing a new understanding of the present and designing new possibilities for the future became vital for social innovation initiatives "not only [to] propose new ways of doing and organising, but typically also [to] engages in the construction of reality through new ways of framing and knowing" (Wittmayer et al., 2019, p. 3).

A social innovation process have its crucial moment in the transformation of a shared imaginary into narratives powerful enough to lead individual and community agency. Its success is proportional to the capacity to mediate collective desires through slogans, stories, symbols and material elements to influence political decisions (Wittmayer et al., 2019). According to Julia Wittmayer et al. (2019), three phases characterise the emergence of these narratives of change: "Firstly, a variety of activities that members of social innovation initiatives engage in contribute to the distributed construction of narratives of change. Secondly, such construction can be done in a more or less hierarchical or deliberative way and thus provide more or less voice to different individuals and initiatives. Thirdly, a plethora of narratives exists at different times and different levels of analysis (i.e. societal discourses, personal stories, organisational myths), which interact and mutually influence one another" (p. 3). Making imaginaries into coherent narratives is a long-lasting process, the result of which is uncertain, and effects are even more so. It is therefore not surprising that a sweeping interpretative flexibility exists with regard to the definition itself.

These considerations allow us to better understand why DSI, being a relatively newly defined phenomenon, is still subject to plural and often contrasting interpretations that affect the aims and processes themselves, explaining the relevance of the different socio-technical imaginaries it is associated with. Bootstrapped by the digital revolution, imaginaries on social innovation today entail a deep integration of technology, materiality and societal transformation. The traditional vision of innovation based on the division between innovators and users has been challenged by the burgeoning literature and mushrooming experiences in bottom-up innovation, user-centred innovation, distributed innovation, community-based or grassroots innovation (Von Hippel, 2005; Seyfang & Haxeltine, 2012). This has been supported by the diffusion of two narratives. First, the narrative of an unprecedented possibility of democratisation that is offered by the diffusion of ICTs, associated with ideals of greater accessibility, transparency and engagement (Charalabidis & Koussouris, 2014). Consequently, technologies not only allow but are also supposed to induce the wide participation of people. Second, the belief in the "wisdom of the crowd" (Surowiecki, 2004), which has had a profound impact in shaping our vision of a collaborative society of expert citizens operating through decentralised and connected platforms. For instance, the whole movement which emerged in the last decades, promoting open-source software and open-access information technology, clearly shows the importance of socio-technological imaginaries in the production of social-economic changes because the democratic mode of co-producing digital innovation is mobilised as a normative social model (Joly, 2019). While, in fact, open-access practices are now common in computing and information technology, "The implications of peer-to-peer go well beyond computer systems, and some scholars predict that in the information age it becomes the basis for a new socio-political constitution" (Joly, 2019, p. 38). The dialogue, prefiguration and reflexivity on democratisation processes leads to an alignment of science and society as a major lever for addressing grand challenges. For instance, the invention of collective property rights through general public license has given a strong impetus to co-production as it allows free use, to modify and distribute innovative software—it therefore conveys an imaginary of broad accessibility not only to the content but also to the production of the web space, the biggest public arena ever created.

Nevertheless, as in any other period of human history, also in the digital age, the adoption or transition from one to another imaginary is accompanied by changes of their supporting narrative, it is not smooth and immune from tensions (Smith & Tidwell, 2016).

4.3 Different Contexts, Alternative Definitions

The collective understanding of DSI has not yet stabilised, neither have the technical imaginaries underlying it. These imaginaries are reflected in the organisational structures, adopted tools and pursued goals of socio-technological projects, working as a prefiguration of innovative forms of social life and social order. As the following analysis makes clear, the heterogeneous domain of social innovation advanced through digital means and processes is still subject to a considerable degree of fuzziness. This heterogeneity leads to a situation in which different definitions emerge concerning the same group of activities; and raises questions about what imaginaries lead different communities to adopt one definition rather than another.

As we have established, DSI is far from an institutionally stabilised and shared concept. The status of interpretative flexibility in which DSI still lays is an understandable result of the different communities of actors, geographical ties, multi-layered practices and culturally specific contexts from which they emerge. Scientific literature has already documented multiple alternative concepts (Rodrigo et al., 2019; Maglavera et al., 2019; Stokes, 2020; Ozman & Gossart, 2019) and their association with different imaginaries and systems of values. Notably, in the European social innovation and policy context, the term DSI is adopted as an affiliation of the broader *Collective Awareness Platforms* cluster (Anania & Passani, 2014) established by the EC since 2014. In other academic, business and institutional contexts alternative names are commonly adopted, including "Civic Tech" (David et al., 2018), "Tech4Good" (Podder et al., 2018) or the more specific "ICT-enabled Social Innovation", which refers to the initiatives of digital improvement for government functions (Misuraca et al., 2017). Less referenced names include "Technology on Civic Life" (Cavanaugh, 2003), "social embeddedness of technology" (Dobryakova & Kotel'nikova, 2015) and similar.

A simple web scraping, performed through open-source search results clustering engine *Carrots2* (which is able to automatically organise internet search results into thematic categories, and clusters the content of up to 400 websites resulting from a web search conducted *via Google Search*) shows the different universe of values, meanings, goals and networks associated with the principal names that populate collective imaginaries (Fig. 4.1).

Scraping results indicate how different definitions are adopted in different domains. The DSI concept is strongly associated with the European context and EU-supported *Research and Innovation* initiatives (keywords Digital Social Innovation for Europe, European Union, project on DSI). Particularly, it shows the existing connections between the digital innovators' communities and digital activists (keywords DSI lab, open or collaborative innovations, DSI fairs) that adopt open-access,

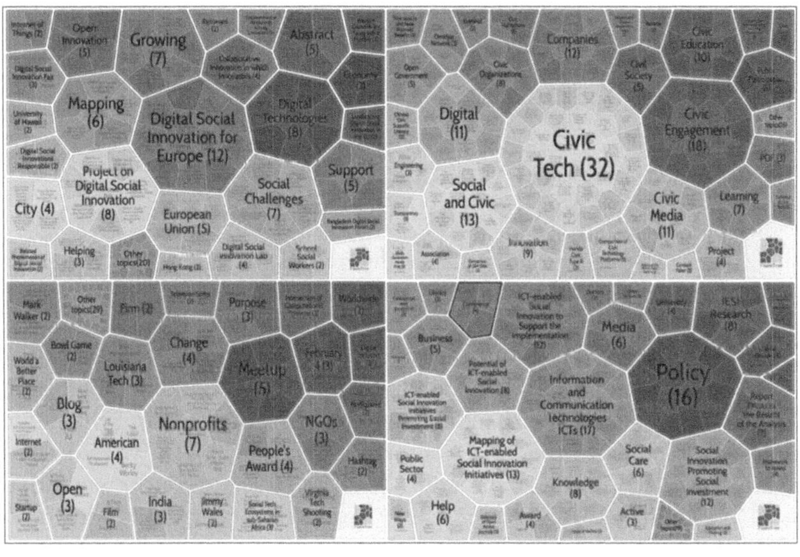

Fig. 4.1 Honeycomb maps of web pages associated with DSI and alternative concepts *via Carrots2* (as per June 2020): top right, DSI; top left, CivicTech; bottom right Tech4Good; bottom left ICT-enable social innovation. General results (denoted from keywords as "abstract", "support" or "helping" and similar) are not considered as relevant

open-source and co-creation approaches for collaborative innovation tackling social challenges. The Civic Tech concept is mainly adopted by private organisations (such as ICT companies, social entrepreneurs and civic associations) and chiefly refers to the facilitation of citizens' participation in e-government (keywords civic engagement, civic media, civic education, civic society). The definition is mainly used in the U.S. context and refers to initiatives that have in time received large support by private ICT companies (see Sect. 3.4). The Tech4Good concept is mainly adopted by civil society, NGOs and in general activities from non-profit organisations, mainly in non-European contexts. Here DSI initiatives are mainly performed through a potentiated use of the social web. Last, initiatives defined as ICT-enabled social innovation focus on social innovation supported by the social web and AI to support the implementation of social welfare measures with the overall aim of impacting on policy. The definition, introduced by the *Joint Research Center of the European Commission*, has been mainly adopted by policymakers and consultants in Europe, inspired by the model of the "enabling welfare state" (BEPA, 2010) and the "socially entrepreneurial state" (Mazzucato, 2013; Misuraca et al., 2017).

4.4 The Alternative Imaginaries of Digital Social Innovation and Civic Tech

The existence of multiple alternative definitions, together with the disciplinary compartmentations we mentioned in Chapter 2 (which often produces unrelated research on the use of digital technologies for social purposes) makes it very difficult to isolate a corpus of relevant literature. Overlapping definitions and the continuously mutable context of DSI associated practices requires deep scrutiny of what is behind the names. It requires considering whether all forms of digitally empowered SI are intended for the same aims, who are the innovators and what social paradigm or worldview do they aspire to, and furthermore, who is benefitting from their results.

From a bibliometric analysis and mapping process on the *Scopus* database (one of the most reputed repositories of high level, international academic literature), performed with the *VOSviewer* software, performed in 2020, it emerged that still by the middle of 2020 contributions were limited to about 140 scientific papers. A graphic representation of the internal articulation of this literature shows the principal research topics that function as attractors of scholarly interest (Fig. 4.2).

The map is very dense and there are several overlaps and interconnected relationships among the clusters that show how interactions between different projects and the promoting communities of practice are very intense. A few trends are detectable, including the challenges faced by individuals and society in the digital democracy (represented by the yellow cluster), the relevance of human factor in information-gathering and processing attention (represented by the blue cluster), civic engagement (represented by the green cluster) and the deployment of collective intelligence (represented by the red cluster) or the open-data tools (represented by the purple cluster). Overall, however, it is clear that the relationship between society and technology (e.g. democracy, community, civic engagement, participation) is vastly the most debated aspects, with comparatively little attention for the more technical issues (e.g. gamification, blockchain, design...). Nevertheless, the timid clusterisation of Fig. 4.2 signals that the domain is not yet mature because it is still subject

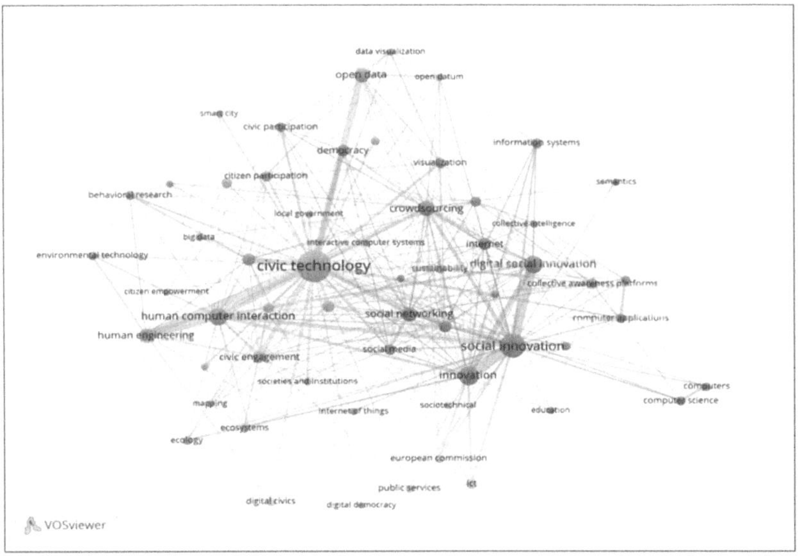

Fig. 4.2 Topics emerging from scientific contributions on DSI and associated definitions (elaboration by Filippo Corsini)

to public experimentation and discussion on how techno-scientific rationality of the digital age and technological tools are transforming social life.

The most apparent categorisation revolves around the two main attractors: one connected with the concept Civic Tech (mentioned in the 50% of the scientific contributions); and a second which gathers together SI, DSI and Collective Awareness Platform concepts (in 54% of the contributions). By digging deep into semantic distinctions, it clearly appears that different concepts have been adopted by different geographical and political macro-communities and economically supported by different sources. These refer to different socio-technical imaginariesof digital agency for the public good, and are utilised in alternative narratives characterised by often contrasting socio-political ideologies and aspirations. Together with gathering different geographical communities (notably the European one for DSI and associated terms and the U.S. one for Civic Tech), these also play the role of attractors for several, interconnected subtopics. For instance, while DSI links with the cluster of ICT and innovation in public services (i.e. the yellow cluster), Civic Tech is connected with issues of civic participation and engagement (i.e. the green cluster). A deeper analysis of this bipolar partition reveals interesting details.

The Civic Tech cluster is principally rooted in research on technologies for government, information and communication. It refers to any technologies intended to deploy the civic virtues of the U.S. traditions *via* opening communication channels between the government and the governed. Not surprisingly, the *Civic Graph* initiative by *Microsoft* mapped existing projects that are defined as Civic Tech initiatives and that are mainly located in the U.S.

In this perspective, technology can help governments to disseminate information to the general public, and for citizens to have their voices heard. It can impact on the different phases of digital democracy and digital participation processes. In fact, the report from the U.S. based *Knights Foundation* on Civic Tech (Patel et al., 2013) clustered existing relevant projects into: open government (data accessibility and transparency, data utility, public decision-making, resident feedbacks) and community action (civic crowdfunding, community organising, crowdsourcing, neighbourhood forums, peer-to-peer). It also signalled that for years to come, we can expect a massive increase of p2p and community organisation proposals.

The dominance in the U.S. context of the private sector as the bootstrapper for civic initiatives (*via* the provision of new technological tools and processes for enhancing citizens' dialogue with the administration) is coupled with the limited intervention of public governments and administrations. On the contrary, in the European digital space, public institutions have a key role in promoting ICT-enabled social services, supporting local entrepreneurship of digital innovators whose commitment in solving social problems can only contemplate the increase of private profit as a side effect.

The DSI (and associated concepts) literature is largely European centric, with a pivotal role of Italy and the UK. Here the relevance of the social innovation attractor in Fig. 4.2 is justified by the recent re-emerged interest for the topic in the age of digital turn (see Chapter 1) supported by EC-funding programmes. At the same time, the theme of human–computer interaction is also central in the map. Several interactive technologies based on user-centred concepts are difficult to operate as well as the instruments used to foster user participation such as blockchain and IoT.

The macro-regional (in this case, the U.S. and the EU) and sectorial (i.e. public institutions, business, third sector) contexts have a profound impact on the material and semiotic construction of a digitally enabled society; and they generate different imaginaries and narratives of the socio-technical entanglement of the digital with social problems.

The socio-technical imaginary associated with Civic Tech recalls an efficient government enabled and improved *via* digital development thanks to the cooperation of private tech companies (which provide the know-how, the infrastructures and financial means) and is open to citizens' collaboration. Social ordering is here determined by the alliance of public and large private institutions. Civic Tech is imagined as able to "offer, at least in theory, a more egalitarian public sphere where community residents can be empowered not as passive recipients of governance, but as active participants in the co-creation of policies" (Russon, 2017, p. 744). In such a narrative and the corresponding political vision, there is no room left for revolutionary bottom-up changes in the neoliberal socio-economic organisation of digital capitalismbecause Civic Tech is essentially proposed as a way to change and make efficient the ways in which bureaucracy relates to citizens.

On the other hand, the imaginary associated with DSI prefigures a proactive and responsive society composed of heterogeneous actors in-between public, private and third sectors (i.e. civil society, digital activists, social entrepreneurs) who are able to apply their knowledge to their contextual conditions, mobilise expertise in order to improve governance (and government) processes *via* digitally mediated life experiences. Therefore, public institutions play a key role in defining the social model by supporting and facilitating the participation of citizens (occasionally in the form of small enterprises, fab labs, start-ups, associations). In this context, the overall goal is to address social issues within a progressive values frame which alternates moderate liberalism and European social-democratic political traditions. This balance is particularly difficult to achieve in a time of austerity constraints. The potentiality of social innovation to tackle major social problems has been particularly well-received because, as the *Bureau of European Policy Advisors* (BEPA) pointed out: "at a time of major budgetary constraints, social innovation is an effective way of responding to social challenges, by mobilising people's creativity to develop solutions and make better use of scarce resources" (2010, p. 7). Expectations are very high. As the white paper of the EU-funded *Chic* project clarifies: "Digital innovation can drive progress towards realising the UN Sustainable Development Goals and shore up the three pillars on which they are built: improving people's quality of life; fostering equitable growth; and protecting the environment. [...] The benefits to the citizens and the society are immense including significantly increased productivity, economic growth and greater employment opportunities". (Maglavera et al., 2019, p. 7). Therefore, the imaginaries of DSI, as a progressive and virtually infinite set of practices, is enriched with further dedicated imaginaries about the technological tools that decentralise power to citizens and communities and accelerate innovations by aligning them with social needs. For instance, "The development of open data, federated identity, bottom-up wireless and sensor networks, open hardware and distributed social networks can potentially serve collective action and awareness" (Bria, 2014, p. 19). However, counterforces are also at play and can "limit the ability to grasp the opportunities that digital transformation for the society presents" (Maglavera et al., 2019, p. 7). These include, for instance, the de-globalisation trend and rising protectionism, inadequate regulation and uneven technology adoption.

The most promising effort performed by a public macro-regional institution, the EC, together with researchers and activists to re-define the rule of digital governance and redesign the internet space, is definitely worth deeper investigation. The domain of DSI itself is not internally homogenous and frictionless, but it encompasses a number of different practices, arrangements, meanings and intentions, imaginaries and narratives. These represent the object of analysis in the rest of this book.

4.5 Unpacking Digital Social Innovation. Functionalist, Reformist and Revolutionary Perspectives

The exploration so far suggests that DSI as a new social technology has still to find a stabilised or shared vision, a common consensus on either the problems to solve or means to do so. At a closer gaze, the world of DSI reveals its internal articulations and tensions determined by the co-existence of diverse approaches emerging in different geo-cultural contexts, adopted by different social actors, inspired by different values and aiming at different goals (Törnberg, 2018).

To disarticulate the dense and quite magmatic ensemble of DSI initiatives, research that recalls the traditional differentiation between multiple social innovation approaches can be of help (Maglavera et al., 2019; Misuraca et al., 2017). Such an approach would need to place DSI somewhere along the imaginary lines between functionalist and revolutionary perspectives: the first, intended to reproduce the status quo and the last, aimed at subverting it. Between the two, there is a vast array of nuanced approaches.

Initiatives intended to make existing socio-political and economic processes faster, more efficient or effective without inducing any significant change in the functional processes supporting the social (re)productionor in its understanding are those defined here as functionalist. The aim of these functionalist DSI processes is to reconfigure the market structures and governance patterns to create social services that meet demands to which the State is unable to respond. This is done by deploying the creativity and entrepreneurial ability of private actors. They often require the elaboration of technological innovations to facilitate automation of tasks and improve the quality and efficiency of internal and external business processes. They entail innovative organisational processes used to support or complement service provisions. It is thus not surprising that functionalist DSI initiatives smoothly integrate into the smart city rationality, and are often supported and even co-opted by neoliberal institutions.

It is clear how Civic Tech initiatives presented in the previous section, often supported by big tech companies, can be easily associated with the functionalist approach. In fact, despite being presented as examples of participatory adoption, they often serve the purpose of advancing the company's business and provide the institution a means to better control society. In the DSI field, which presents itself as the counterpart of the U.S. liberal ideology of market-dominated innovation, it might be trickier to identify the functionalist approach in action. However, as some authors recognise, we can detect a file rouge connecting the two shores of the Atlantic: "We can see this in the Young Foundation and NESTA's championing of distinctly entrepreneurial conceptions in the UK, spearheaded by their (former) Chief Executive and New Labour policy guru Geoff Mulgan (2007); in the US where Obama instituted a Social Innovation Fund and an Office of Social Innovation within the White House [...]; and its incorporation into policymaking in Quebec, Latin America and especially the EU" (Thompson, 2019, p. 1174). In fact, in the context of neoliberal capitalism, the availability of digital technologies (including those intended to address societal needs) aligns with a business-oriented understanding of innovation (Dacin et al., 2011). DSI is thus described as a possibility for supporting new forms of micro-privations of services and light infrastructural provision. Their social aspects often reside in providing an occasion for entrepreneurs to promote a distributed model of social and economic capital production, with marginal attention towards the social critiques. The very nature of DSI positioned between public and private domains fits perfectly with this approach; it determines the co-optation of DSI initiatives in the neoliberalist vision of "everybody-entrepreneur" (Stokes, 2020). In practice, this feeds a new social order where start-ups, micro-enterprises and big economy jobs only create social mobility. In fact, they often generate new forms of precarity and of labour force exploitation. The functionalist approach, thus, often (but not always) benefits Big Tech capitalism (Keanbirch, 2019). It contributes to eroding the authority of public institutions. These public institutions are rarely able to integrate technological changes in their operational functions at the same fast pace of their production. This means the power of intermediary bodies crumble for the very reasons why there are often no intermediary bodies to represent the emerging social groups, thus nullifying the possibilities of the state and macro-regional organisations to act as guarantors of social equality and justice.

Proceeding along the imaginary line between functionalist and revolutionary initiatives, we can identify a reformist approach that characterises many DSI initiatives as more inclusive, participatory and people-friendly than functionalist ones. These are advanced by CSOs and public administrations to promote the collaborative platforms where citizens can enter into dialogue with public administrations and with an ideally non-predatory private sector (Fig. 4.3).

By directly intervening and transforming the functioning of socio-spatial structures, reformist initiatives deploy the potential of existing digital technologies and leverage their co-creative and pervasive capacity to reform institutions. For instance, initiatives aimed at helping migrant people emerge from invisibility by learning local languages (e.g. *Speak*) and at smoothing the creation of mutual aid networks in the city (e.g.

Fig. 4.3 An expert consultation process in Ghent City Council Hall, 2018 (photo by the author)

Citizen Foundation) are representative of a reformist approach. While some use proprietorial technologies (e.g. *Fixmystreet*), others couple their social commitment with concerns for the ethical aspects of the digital and adopt open-access and open-source technologies (e.g. *Smart Citizen*).

On the opposite side, we find revolutionary DSI initiatives advanced by cyber activists, hackers and e-makers with the explicit intention to modify rules and tools of the digital world. By following up one of the initial intuitions of the fathers of digital revolutions, i.e. that we can only change society by changing the tools it uses (Cadwalladr, 2013), they invent, hack, boycott and transform the technologies of everyday life, to contrast the massive fluxes of the economic, financial, material and symbolic powers of digital capitalism (Coleman, 2014; Zuboff, 2019). They fight to promote people's digital sovereignty over their data (e.g. *Lleialtat*), access to the web through community-owned network infrastructures (e.g. *Guifi*), or the possibility for self-fabrications of technological tools (e.g. *Arduino*). These initiatives promote innovation that happens outside institutional settings and strive to reconceptualise the problems they themselves are called to solve. They do so by intervening on cognitive frames of reference (Nicholls et al., 2015), and through the elaboration of alternative interpretations of the role of technologies in the constitution and reproduction of society.

Nevertheless, as with every social phenomenon, DSI and similar processes have fuzzy characters and, in some cases, the revolutionary practices align with the neoliberal paradigm of digital capitalism(Bendiek et al., 2019; Parayil, 2005). As Wittmayer et al. (2019) clearly explain, we cannot forget that the notion of social innovation has been adopted in the past years to broaden the technology-oriented paradigm of innovation (e.g. the *Vienna* Declaration (2011) and the *Lisbon* Declaration (2018)). It has been intimately connected with "the master narrative of innovation for growth and its related socio-technical imaginaries" in Europe (Strand et al., 2018, p. 1850). Considerations on the different forms of DSI need to be moderated by the attention for the sometimes-ambiguous relationships between social innovation and neoliberalism, where the first is put into service of the second (Moulaert & Van Dyck, 2013; Swyngedouw, 2005) and fuels processes of depoliticisation often to justify neoliberal public policies (Fougère et al., 2017; Jessop et al., 2013; Schubert, 2018).

In this vein, it is important to acknowledge that functionalist DSI initiatives do not focus on topics that are significantly different from those dealt with by more revolutionary ones. Key differences reside in the management process and the final aims. Notably, the two approaches present a different understanding of the role of public and private actors in the governance of urban life. While the first promotes digital governance processes supported by the private investments of ICT companies, the second envisages and regulates DSI processes on the basis of the assumption that the economic profit of DSI entrepreneurs is only intended as a side effect of social problem resolution (Nicholls et al., 2015).

From one side, functionalist initiatives entail the mobilisation of companies through privatisation of public services and welfare systems, the devolution of public competences (including the policy-decision-making, financial and economic planning, development plans, etc.) to public–private companies, the adoption of proprietary digital facilities for hosting public debate or similar, with the aim of prioritising business development and private capital expansion (Betancourt, 2016). From the other side, revolutionary DSI is understood as an attempt at contesting and re-making society and its functioning without serving the intents of the neoliberal governmentality. It claims to pursue the ideal aspiration of expanding social (rather than private) capital; and question the visions, narratives and imaginaries that materialise in technological innovation, along with the hidden disciplinary power of ICT companies in the formation of political agendas (Zuboff, 2019). Contrary to the functionalist approach, the revolutionary one requires inventing and adopting a horizontal model of digital innovation management, at the local scale. For instance, the creation of independent, city-owned federated web sites in collaboration with digital activists for developing in-house competencies, publicly owned infrastructures (e.g. *Hallo.gent* platform developed by the City Administration, the independent company Ind.ie, and the public-private *Digipolis* Institute in Ghent) can transfer the ownership of internet platforms to public administration.

Similar DSI initiatives advance socially progressive aims by co-designing the "internet of humans" (Viola, 2020) "that's egalitarian, decentralised and free of snooping" (Harris, 2018), as the only viable alternative to the neoliberal technological development (Söderström et al., 2014).

When innovation is decoupled from digital capitalism, it can be re-appropriated for producing emancipatory social changes (Avelino et al., 2017). By following this inspiration, the revolutionary approach explores what is beyond the critique of capitalism. Far from the venture capitalism of big tech business models with their fake all-free-for-all mantra,

DSI initiatives can reinforce the commons (e.g. *P2P Foundation*); revitalise public institutions and reinvent redistributive measures (e.g. taxation systems that impact on big capital, the social policy consultant network promoted by the *New Economic Foundation*) and elaborate co-creative technologies for the public benefit (e.g. the *Arduino* open-source electronic prototyping platform to create interactive electronic objects).

The EU project DSI4EU embarked upon describing the contrasting imaginaries around the power of digital technologies in social life and around the opportunities to engage with digital innovation initiatives, be that as a proponent or simple user. They clearly pointed towards challenges and means to address them: "A major risk for the *Future Internet* is the realisation of the 'Big Brother' scenario, with big industrial players (mainly US-based) reinforcing their dominant position by implementing platform lock-in strategies, enforcing extensions of copyright and patents, appropriating users' data and discriminating network traffic. By centralising computing, data storage and service provision (*via* the Cloud), and by striking strategic alliances between the largest Over-The-Top (OTT) and largest network operators, there is a risk that the innovation ecosystem will become more closed, favouring incumbents and dominant players, thereby in time constraining user-driven innovations, particularly ones that don't involve monetary payment" (Bria, 2014, p. 19). This trend in the digital sphere, of centralised profit-based monopolies, is in contrast to "innovation activities and research projects built peer-to-peer and small scale local social media, building on small community networks. This is a bottom-up explore-as-you-go experimental approach" (Anania & Passani, 2014, p. 4). The potential of revolutionary DSI to work as a form of social critique is explored in Chapter 5.

4.6 When Digital Social Innovation Meets Urban Imaginaries

The existence of alternative definition of digital technology-based practices to tackle social problems in different macro-regional contexts illustrates that DSI has strong territorial tights. The physical, geographical space, with distinctive proximity linkages, infrastructures and environmental conditions have an impact on the practical performance of DSI initiatives that generate specific understandings of space (the digital spatialities mentioned in Chapter 2) and allow different possibilities of intervention. Also, the cultural communities they are embedded within have a significant impact in determining what innovators can reasonably achieve, what kind of support they can get from institutions and businesses, whether their initiative is getting socially accepted or not. The urban spatiality (i.e. the imagined, perceived, experienced space of a city) plays an important role in this process. In fact, in their various forms, DSI practices embody key tenets of the digital revolution that are also reflected in contemporary city governance:

- Disintermediation: To react to global challenges, cities directly connect with global actors (being international institutions, multinational companies, civil society global movements) without intermediaries, to advance appropriate strategic changes. For instance, the role of urban governments in producing effective policy strategies to fight climate change is well exemplified by alliances of cities, such as *EuroCities*.
- Participation: This is not a new trait. Participatory traditions have solid roots in urban governance, but the possibilities offered by the social web opened new opportunities for massive data collection and elaboration. Linked with issues of democratisation and engagement, participation is the mantra in contemporary urban agendas.
- Dematerialisation: Most of urban social life is today translated into the augmented urban space (see Chapter 2) which is constituted by social media and interactive web platforms. These last allow cities to increase their economic solidity without (necessarily) physically expanding or investing in hard infrastructures for production, through the expanding service sector (including logistics and tourism).

The imaginary of the contemporary city is deeply affected by the promises of the digital revolution. Globally connected hubs, open to innovative experiments at the frontier of socio-technological hybridisation which are inclusive, engaging, socially cohesive and environmentally responsible, economically solid, politically vibrant while being reliable, trustable, enthusiastic and welcoming. Who wouldn't live in a city like this? However, analysing the city's relationship with DSI processes can offer a new, more complex perspective.

Urban imaginaries have power on the elaboration of policy provisions, normative frameworks and technical infrastructures that are able to deal with the fluid character of emerging processes of public governance, where competences, responsibilities and power circulate through the network of social actors, by giving raise to mutable (and sometimes unexpected) socio-technological assemblages. DSI and contemporary urban imaginaries combine, deeply affecting the integration of digital technology into the day-to-day functioning of the city. These imaginaries are "crystallised" on artefacts and cultural products (notably in discursive forms) and unveil the conflicting forces, meanings and visions behind the multiple narratives of urban development.

Three imaginaries are most frequently proposed. The first, and most prolific, directly emerges from the smart city programme and can be named as the hyperconnected-city imaginary. The second, which can be labelled as the receptive-city, is characterised by mitigation measures that counterbalance the technology-enthusiasm with social concerns and envisages the regulatory intervention of public authority. The third roots in the digital anarchist tradition and while not necessarily taking extreme subversive positions, describes cities as the outbreaks for the digitally empowered social revolution to come. It can be defined as the do-it-yourself (DIY) city imaginary.

4.6.1 The Hyperconnected-City

The hyperconnected-city imaginary offspings from the smart city paradigm with a central role attributed to the technological readiness in generating digital solutions to urban management problems (Gladwell, 2013; Turkle, 2011). Following the vision of the ICT companies that established the smart city paradigm, such as *IBM* and *CISCO*, digital interconnected technologies are meant to help "cities run more efficiently, save money and resources, and improve the quality of life for citizens" (IBM, 2021). This perspective reduces most of the complex problems faced by the contemporary city to an issue of efficient control and direction. The resulting hyperconnected-city is, thus, portrayed as an incubator of efficient and accountable organisational processes, facilitated or even driven by digital technologies and able to involve citizens in urban management tasks. The underlying assumption justifying the need for building up an increasingly digital city is that broadening technological access (both in terms of availability of technology and literacy) will lead to greater citizen participation. In turn, citizen participation is regarded as an opportunity for developing new software to make the most of the urban hardware (Dyer et al., 2017), supporting a virtuous circle of economic growth.

For instance, e-government initiatives, e-voting and crowdsourcing platforms are described as entering gates for lay citizens in the policy-decision process. They can contribute by providing their opinion about political parties' electoral programmes or legislative initiatives (e.g. the *Citizens' Initiative Act*, designed by *Open Ministry*, a non-profit organisation based in Helsinki).

However, the practical enactment of the hyperconnected-city imaginary raises doubts and criticisms. Notably the definitions of the goals of the relevant initiatives are questioned as pre-given by the promoters (whose final aim could be driven from economic or reputational interest, rather than from the collective well-being). Again, citizens' roles in most of the initiatives contributing to the hyperconnected-city are often limited to passive providers of information or data. Exemplary crowdsourcing platforms such as *MK:Smart* promoted by Milton Keynes administration or the Swiss *Citying platform* created by a private company, raise questions on both the very understanding of the meaning of participatory practices and on the intents of the initiatives themselves. When digital technologies, particularly those embodied in personal devices, pervade and direct individual agency (by defining the limits of what can and cannot be done), it is important to ask whether the intentionality of the process is defined by the user or by the tool (which embeds the designer's aims, the technical, material or logistic constraints, the inter-connectivity requirement of the IoT, as well as the particular logics of AI). In many circumstances citizens function as bodily "extensions" of the algorithms they use; as "living sensors" for recording data about air quality, traffic and mobility, consumption behaviours and similar; and as players in games somebody else invented and gave the rules to. It is, in fact, not technological readiness per se, but rather the systemic reception and upscaling of innovation that marks the difference in cities' capability of mobilising their relationship with citizens and their knowledge of local needs (Barcelona Activa, 2018, p. 11) to fulfilling the promises of the technological citizenry.

Despite the values of public participation, transparency and openness referred to in hyperconnected-city plans, the corporates' storytelling of a smart, hyperconnected-city (Söderström et al., 2014) prove to be surprisingly resonant with the policy programmes of many public administrations, whose digitalisation plans are entirely managed by private companies. Therefore, the power of ICT service providers in the creation of the urban future vision and the definition of the agenda of interventions has drawn criticism that the hyperconnected-city is a production of commercial enterprise imaginations, rather than of citizens (Cocchia, 2014; Townsend, 2013). For instance, critical scholars have documented how the rhetoric of the hyperconnected-city utopia, promoted by many of those enthusiasts of the smart city programme (Vanolo, 2013), are unveiling the dystopian effects of a Silicon Valley-dominated pathway to technological development. A flourishing debate discusses what the real (Kaika, 2017), truly (McLaren et al., 2015) or actual the smart city is, and how to produce radically transformative programmes (Hollands, 2015; McFarlane & Söderström, 2017) which avoid technological fetishism of the hyperconnected imaginary, and also eschew passive and tokenistic participation risks (Cardullo and Kitchin 2019; March & Ribera-Fumaz, 2014), while producing alternative bottom-linked governance processes (Leitheiser & Folmann, 2019).

4.6.2 The Receptive-City

The imaginary of a receptive-city is adopted in many initiatives proposed by national or local public institutions and it gathers together public and private actors who mitigate the hyper-technological fantasies with more cautious strategies for creating a people-oriented city. Among the initiatives intended to empower people beyond attracting their participation, *PublicLab* was created by a citizens' association in 2010 in the wake of the *British Petroleum* oil disaster to help people track local effects of environmental damages and to guarantee their right to access knowledge. Another example is *CitizenLab*, a digital participation platform offered by a Brussel-based company to worldwide governments (already adopted in Denmark, Belgium, Germany and France) that aims to get a user-friendly tool to consult citizens and include them in decision-making. A further example of how to collect ideas on urban development plans is the platform *Otwarta Warszawa* created by the small IT company, *MillionYou*, in collaboration with the Warsaw city administration. It decides the final implementation of emerged proposals. A similar project is *OurMK* promoted by Milton Keynes local government in cooperation with The Open University. With the declared aim of accelerating a transition to sustainability, the *OurMK* programmes calls for digital problem-solving processes to address real-life complex issues, and offers financial and logistic support towards the realisation of winning proposals. A further significant example is the *Better Reykjavik* platform connecting Reykjavik city council with citizens, whose main problems are evaluated by the municipal offices and assigned a budget for their tackling. The imaginary of a city able to listen and give adequate consideration to everybody's needs, underlies all of these projects and is characterised by a strong commitment of the public institution in jointly, autonomously promoting projects that can fuel the dialogue between different sectors of society. Along this line, it is equally common to find public administrations or civil society organisations organising participatory planning contests in which citizens collaborate with medium or small enterprises to support the management of urban services. An example of this is the *Prize2theFuture* contest, implemented by the Community Foundation of Greater Birmingham and developed by the *Idea Crossing* company, which was intended to gather ideas for transforming Birmingham "into a cooler, more vibrant city" (Prize2theFuture, 2021). Leveraging on crowdsourcing processes, these contests are often included in participatory budget programmes, as in the case of the *BurgerBudget* initiative in the city of Ghent (see Chapter 1). Similarly, the *Paris Idée* platform was created by the city council to harvest ideas for a participatory budget, allocating 100 million euros to winning projects. Crowdsourcing platforms are slightly different, used as consultation sites where citizens have the opportunity to express their opinions on already existing (and sometimes contested) urban development projects.

After some years of experimental applications, the path towards the digital empowerment of citizens and ICT facilities-based governance processes still seems uphill, with many revolutionary wishes doomed to remain un-realised (or even un-realisable); and other wishes turning out to be probably not so desirable, due to the intrinsic difficulty to govern the digital sphere and the emergence of potential and unintended socially regressive outcomes (see Chapter 5). While DSI processes have the possibility to empower people and let them responsibly take part in the public life of the augmented city; the interference of ICT business corporations and financial institutions into urban governance raises concerns about the actual freedom, transparency and ethical motivations behind digitalisation. It is therefore necessary to ponder this potentially contradictory dynamic if the full implications of digital innovation for participatory governance are to be acknowledged.

TextBox 4.1. What a Receptive-City Looks Like. The Case of Ghent

Ghent is a dynamic mid-sized city in the centre of the Flanders region of Belgium. It is characterised by the inhabitants' long-lasting commitment to public participation (Bauwens & Onzia, 2017) and a strong attitude in pioneering innovative solutions to increase social cohesion (Baccarne et al., 2014). In this "city of troublemakers" (Tourist Office Ghent, 2019), strategic planning is supported by remarkable stability in the city administration of left-side political alliances. The tight relationship between the city administration and citizens is backed upon a mutualistic and commons-oriented tradition; and justifies the co-optation of activists and scholars in the public administration business, together with the presence of civil servants in the city neighbourhoods (Sharable, 2018). While some activists point out the distortive effects of the professionalisation of participation, they also recognise that this makes it possible to combine social antagonism and institutional perspectives in an effective way. This progressive character of Ghent is evident in the use of joint platforms for associations, intermediary organisations and private entrepreneurs that provide public services and care for the needs of specific social groups (e.g. *Ghent Smart City, City of People, Ghent Living Lab*). Hard and soft urban infrastructures are combined in international (e.g. *EuroCities; Refill*) or locally funded participative projects (e.g. *RuimtevoorGent, GentSintPieters, Ideeen voor MMM, Crowdfounding Gent, Dienst Beleidsparticipatie).*

Ghent's propensity towards risk-taking and experimenting in social participation initiatives, inspired by the intense presence of digital entrepreneurs and social innovators, start-ups and co-makers' initiatives (e.g. *Timelab, Peerby, Manouvre*), attracted resources from the academia (e.g. *Stads Academie*), innovations institutes (e.g. *IMEC*) and the European Commission (e.g. *Interactive Cities, Eltis, REGent* projects) on, for instance, open data (e.g. *Apps for Ghent, De Community*) or urban commons (e.g. *Commons, Commons Transition Plan*). Among DSI projects, the platform *Mijndigitaalideevoorgent* [My digital idea for Ghent] and the Ghent Living-Lab project (jointly funded by the city administration, the *Alcatel-Lucent Bell* company, the EC and the University) focused on how ICT can improve life experience in Ghent (Mechant et al., 2011). A further multi-actorial project was the *BurgerBudget* that devoted part of the city budget to projects proposed, voted and realised by lay citizens (Van Noordt et al., 2018). Larger projects are complemented with digital education initiatives that have been recently set up, for example: digital talent points in the city, digital lending services, annual digital week and interactive fairs, a new public library with a digital section and digital media programmes for mentally challenged people.

People's participation in Ghent digital governance has been investigated by the EC-funded Marie Skłodowska-Curie project 2017–2019 *Crowdsourcing Urban Sustainability Governance. Exploring innovative governance models for addressing urban sustainability through ICT-people interaction* (Fig. 4.4) This aimed to appreciate whether and how crowdsourcing is appropriate for dealing with the mutable, multi-scalar and complex character of urban sustainability, and for empowering technological agency of heterogeneous actor-networks and narrowing the gap between research results and policy applications.

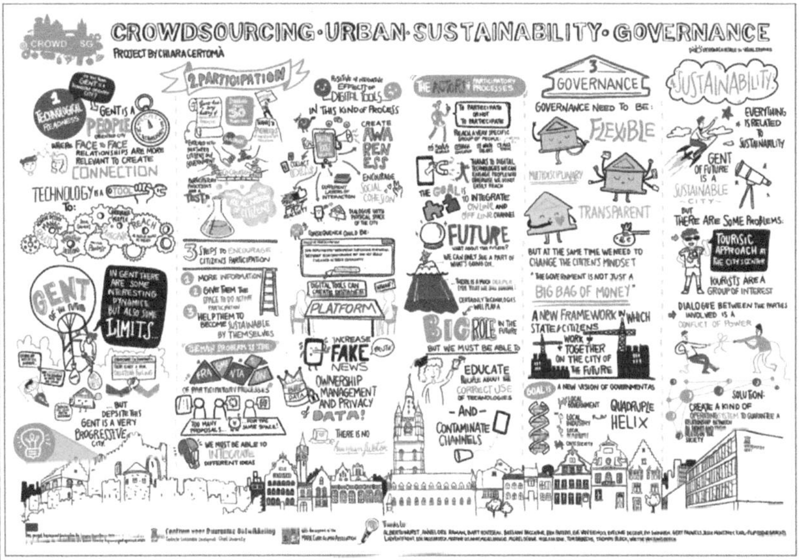

Fig. 4.4 Graphic representation of key tenets of people-ICT interactions in Ghent, EU MSCA project CROWD_USG (text by C. Certomà, graphic by V. Vitale, 2018; full visual story available at crowd.usg)

4.6.3 *The Do-It-Yourself City*

When entering the ephemeral and fast-changing world of digital innovation "imagination of desirable and undesirable futures correlate, tacitly or explicitly, with the obverse – [the] fears of harms that might be incurred through invention and innovation, or of course the failure to innovate" (Jasanoff, 2015a, p. 6).

The dividing line between utopia and dystopia becomes incredibly thin. We can easily shift from the optimistic vision of digital processes able to create an allegedly more cohesive and sustainable hyperconnected-city; through the acknowledgement of existing contrasting forces of public and private interests that undermine the solidity of the receptive-city; up to the critical perspective of the DIY-city. This last aspect reverses the opaquely positive imaginary of the hyperconnected-city into Huxleyan prefigurations of a society of cyber-surveillance, where people's ideas, energy and imagination are limited, co-opted and exploited for private companies' gains and social dissent is pigeonholed and repressed.

Inspired by the digital antagonism and social hacking movement, the DIY-city imaginary builds upon the proliferation of spontaneous processes that link with sharing economy, commoning initiatives and informal spatial planning movements,[1] which adopt co-creation, peer-to-peers tools and processes to support citizens' activism. Advancing DSI initiatives that are grassroots and socially revolutionary, requires the technical competences to hack the market-based infrastructure of the www and turn them into tools for change-making. They need more than a strong political commitment to contest the technological fetishism (Batty et al., 2013; Kaika and Swyngedouw, 2000) of the hyperconnected-city and its social order (Townsend, 2013). The DIY-city imaginaries are fed by revolutionary DSI processes (see Chapter 5 for an extensive description). Originally emerging from isolated experiments, they are now giving

[1] Examples of sharing or circular economy from the grassroots are *Peerby*, a dutch platforms for everybody to sell and buy second-hand products or *Freecycle*, a volunteers' network counting about 9,000,000 members worldwide giving (and getting) stuff for free in their own towns. Common-based and spatial planning movements examples are the *CommonsGent*, a wiki-repository of overly 500 commons-based project in the city of Ghent, connected with a broader investigative project jointly conducted by the *P2P Foundations* and the local administration and *Impossible Living*, a crowdsourcing platform intended to map and reinvent potential use for abandoned buildings.

rise to a large (despite often elusive and non-vocal) decentralised movement. These loosely structured actions rooted in everyday practices are characterised by the coordination of different actors that cooperate for achieving collective benefits and most significantly, are endowed with the intention of contrasting the progressive erosion (Corsin Jiménez, 2013; Hou, 2010), the commodification and privatisation of urban public space (Brenner & Theodore, 2005), and public infrastructures and services, including the digital ones.

When practised with revolutionary aims and through independent means, DSI processes can help innovator communities to understand their potentiality in producing progressive social transformation and in coping with the unthinkable complexity of the digital society by transforming "institutions, overthrowing oppressive 'structures with power', [addressing] non-satisfied needs, [and] building of empowering social relations from the bottom-up" (Moulart & Van Dyck, 2013).

While the functionalist form of DSI contributes to feeding an imaginary of hyperconnected-cities, the reformist approach is resonant in the receptive-city imaginary, and eventually, a revolutionary approach to DSI fits well within the imaginary of DIY-city (Table 4.1).

Table 4.1 DSI approaches and Urban-technology imaginaries

DSI approaches		
Functionalist	*Reformist*	*Revolutionary*
Top down	Mixed forms	Grassroots
Business driven	Government-driven	People-driven
Centralised	Central control, decentralised management	Decentralised
Proprietary	Owned by Institutions	Open
Efficient	Inclusive	Engaging
Market-led	Socially concerned	Politically engaged
Made by a few for the market	Made by the public for the many	Made by the many for the many
Held by companies	Held by the public	Shared
Participation	Collaboration	Empowerment
Hyperconnected	*Receptive*	*Do-It-Yourself*
Urban-technology imaginaries		

As with any taxonomy, this partially obfuscates the complexities and tensions that emerge in complex social phenomena. For instance, the institutional reformist initiative, launched by the EC under the title "Internet of Humans", has embedded, since 2016, large investment on the redesign of technological infrastructures with the Next Generation Internet programme. It is intended "to shape the future internet as an interoperable platform ecosystem that embodies the values that Europe holds dear: openness, inclusivity, transparency, privacy, cooperation, and protection of data" (EC, 2020). These general values have been curved towards a more socially engaged perspective by the DSI innovators communities who have participated in the elaboration of the *Digital Social Innovation Manifesto* (ChiC, 2017). This marked the need to avoid a situation in which the digital agency of European citizens is locked into proprietary systems; the need to build up decentralised internet infrastructures that are able to effectively support democracy and guarantee citizens' sovereignty over their digital life and the need for experimenting long-term solutions that only rely on commercial mechanisms or forms of voluntary participation, but also redesign the rules of internet governance.

The effort devoted by public macro-regional institutions in Europe, to build up a more human-centred internet (see Chapter 5), even from a more radical perspective, has been accused of funding start-ups and technological innovation processes that if successful, despite being paid for with public money are acquired by the U.S. big techs (Balkan, 2019a). Aral Balkan, cyborg rights activist and designer, suggests that alternatives exist for the European public investments in DSI to free our society from these traps: "Let's instead invest in many small and independent not-for-profit organisations and task them with building the ethical alternatives. Let's get them to compete with each other while doing so. Let's take what we know works from Silicon Valley (small organisations working iteratively, competing, and failing fast) and remove what is toxic: venture capital, exponential growth, and exits" (Balkan, 2019b). Disruptive but ethical and critical forms of DSI could suggest effective strategies for social change that can cope with the unthinkable complexity of the digital society.

The analysis of socio-technical imaginaries on DSI and the city started to show how, while possibly being functional to the reproduction of the neoliberal system, DSI can also work as transformative paradigm.

4.7 From Imaginaries to Production

We already noticed that digital technologies, while equipping society to tackle shared concerns, also works as a social technology. It is therefore important to understanding how alternative narratives and visions of DSI, as a technology of society normalisation, control and direction, are mobilised and how criticalities and critical issues they generate are faced.

Whereas Chapter 1 described the roots and meanings of DSI and Chapter 2 made the connection of this to the urban context explicit, this chapter engaged with the initial line of critical geography inspired inquiry, the concept of representation. DSI, in fact, is a social technology whose theory is yet to stabilise, and its meaning is still subject to significant interpretative flexibility. Despite being adopted in reference to multiple applications of digital technologies, the socio-technical imaginariesassociated with DSI have not been explored per se. Such an investigation is essential to understand who is producing the socio-technical imaginaries we build up our collective life upon, which communities are feeding specific imaginaries and for what purposes. Therefore, the above paragraphs explained how the meaning, nature and characteristics of DSI are still in negotiation and how alternative competing definitions are adopted in different socio-political and economic macro-regional contexts and communities of practice. These made evident that multiple perspectives co-exist, from the approach which is functional to the reproduction of the current socio-economic order, to the more reformist and even the revolutionary ones. The emerging socio-technical imaginaries elaborated through distinctive narratives and visions of the future hold a tight connection with the imaginary of contemporary urbanity—characterised by the pervasive presence of technologies. We, therefore, illuminated a link between the alternative approaches to DSI and the alternative imaginaries of the city (i.e. the hyperconnected, receptive and do-it-yourself).

These categorisations raise some issues. As with all socio-technical imaginaries, those connected with DSI and urban technologies are performative and can be operationalised through governance processes. If "imaginaries gain traction and are complemented or strengthened through acts of power" (Delina, 2017, p. 49) it is unavoidable to question how these materialise into physical technologies generated, owned, managed or appropriated in different contexts. The analysis of the urban governance processes entangled with DSI processes proposed in the next chapter investigates the impacts of different socio-technological imaginaries on the city; how these translate into policies, programmes, visions and organisational forms; how these imaginaries are contested or negotiated by diverse communities and how relations of force are enacted—for instance, when economically stable elites displace imaginaries produced by or circulating among less wealthy or powerful groups.

By unveiling underlying governance mechanisms, we can deconstruct the social reproduction processes happening through the digital; unpack the emancipatory imaginary of digital innovation and the dark side of the hyperconnected-city and highlight the opacity of diverse forms of DSI to understand what these aim at and how. On this base, critical DSI research can contribute to appreciating how digital innovations are working as social technologies; and it can allow innovators to understand their potentially revolutionary role in producing emancipatory social transformations.

References

Anania, L., & Passani, A. (2014). A Hitchiker's guide to digital social innovation, 20th ITS Biennial Conference, Rio de Janeiro, Brazil, 30 November–03 December 2014: The Net and the Internet—Emerging Markets and Policies, International Telecommunications Society (ITS), Rio de Janeiro.

Avelino, F., Wittmayer, J. M., Pel, B., Weaver, P., Dumitru, A., Haxeltine, A., Kemp, R., Jørgensen, M.S., Bauler, T., Ruijsink, S., & O'Riordan, T. (2017). Transformative social innovation and (dis)empowerment. *Forthcoming in Technological Forecasting and Social Change*. http://www.sciencedirect.com/science/article/pii/S0040162517305802.

Baccarne, B., Mechant, P., Schuurman, D., Colpaert, P., & De Marez, L. (2014). Urban socio-technical innovations with and by citizens. *Interdisciplinary Studies Journal, 3*(4), 143–156.

Balkan, A. (2019a). *Have you heard about Silicon Valley's unpaid research and development department? It's called the EU*. https://ar.al/2019/06/22/have-you-heard-about-silicon-valleys-unpaid-research-and-development-department-its-called-the-eu/.

Balkan, A. (2019b). *Slavery 2.0 and how to avoid it: A practical guide for cyborg*. https://ar.al/2019/05/02/slavery-2.0-and-how-to-avoid-it-a-practical-guide-for-cyborgs/.

Barcelona Activa. (2018). *Mapping DSI—cities and urban development*. https://digitalsocial.eu/images/upload/33-Cities%20and%20urban%20development.pdf.

Batty, M., Manley, E., Milton, R., & Reades, J. (2013). Smart London. In S. Bell, J. Paskins (Eds.), *Imagining the future city*. Ubiquity Press.

Bauwens, M., & Onzia, Y. (2017). Commons transition plan for the city of Ghent. *Commons Transition*. http://commonstransition.org/commons-transition-plan-city-ghent/. Last Accessed 11 October 2019.

Bendiek, A., Godehardt, N., & Schulze, D. (2019). The age of digital geopolitics. *International Politics and Society*. https://www.ips-journal.eu/in-focus/chinas-new-power/article/show/the-age-of-digital-geopolitics-3593/.

BEPA. (2010). *Empowering people, driving change: Social innovation in the European Union*. European Union, Brussels. http://net4society.eu/_media/Social_innovation_europe.pdf.

Betancourt, M. (2016). *The critique of digital capitalism: An analysis of the political economy of digital culture and technology*. Punctum Books.

Brenner, N., & Theodore, N. (2005). Neoliberalism and the urban condition. *City, 9*(1), 101–107.

Bria F. (2014). *Digital social innovation*. DSI4EU project Interim report, Nesta - EU H2020. https://waag.org/sites/waag/files/media/publicaties/dsi-report-complete-lr.pdf.

Cadwalladr, C. (2013). Stewart Brand's Whole Earth Catalog, the book that changed the world. *The Guardian*. https://www.theguardian.com/books/2013/may/05/stewart-brand-whole-earth-cataloge.

Cardullo, P., & Kitchin, R. (2019). Being a 'citizen' in the smart city: Up and down the scaffold of smart citizen participation in Dublin, Ireland. *GeoJournal, 84*(1), 1–13.

Cavanaugh, J. W. (2003). E-democracy: Thinking about the impact of technology on civic life. *National Civic Review, 89*(3), 229–234.

Charalabidis, Y., & Koussouris, S. (2014). *Empowering open and collaborative governance: Technologies and methods for online citizen engagement in public policy making*. Springer-Verlag.

Cocchia, A. (2014). Smart and digital city: A systematic literature review. In R. Cocchia, P. Dameri, & C. Rosenthal-Sabroux (Eds.), *Smart city*. Springer.

Coleman, G. (2014). *Hacker, Hoaxer, Whistleblower, Spy: The story of anonymous*. Verso.

Corsín Jiménez, A. (2013). The prototype: More than many and less than one. *Journal of Cultural Economy, 7*(4).

Dacin, M. T., Dacin, P. A., & Tracey, P. (2011). Social entrepreneurship: A critique and future directions. *Organization Science, 22*, 1203–1213.

David, N., McNutt, J. G., & Justice, J. B. (2018). Smart cities, transparency, civic technology and reinventing government. In M. Rodríguez Bolívar (Ed.), *Smart technologies for smart governments. Public Administration and Information Technology*. Springer 24.

Delina, L. L. (2017). Whose and what futures? Navigating the contested coproduction of Thailand's energy sociotechnical imaginaries. *Energy Research & Social Science, 35*, 48–56.

Dobryakova, M., & Kotel'nikova, Z. (2015). Social embeddedness of technology: Prospective research areas. *Foresight and STI Governance, 9*(1), 6–19.

Dyer, M., Gleeson D., & Grey, T. (2017). Framework for collaborative urbanism. In C. Certomà, M. Dyer, L., Pocatilu, & F. Rizzi (Eds.), *Citizen empowerment and innovation in the data-rich city*. Springer Tracts in Civil Engineering. Springer.

EC. (2020). *Next generation internet initiative*. https://ec.europa.eu/digital-single-market/en/next-generation-internet-initiative.

Fougère, M., Segercrant, B., & Seeck, H. (2017). A critical reading of the European Union's social innovation policy discourse: (Re)legitimizing neoliberalism. *Organization, 24*(6), 819–843.

Gladwell, M. (2013). *David and Goliath, underdogs, misfits, and the art of battling giants*. Brown and Company.

Harris, J. (2018, February 1). The punk-rock Internet. *The Guardian*. https://www.theguardian.com/technology/2018/feb/01/punk-rock-Internetdiy-rebels-working-replace-tech-giants-snooperscharter.

Hollands, R. G. (2015). Critical interventions into the corporate smart city. *Cambridge Journal of Regions, Economy and Society, 8*(1), 61–77.

Hou, J. (2010). *Insurgent public space: Guerrilla urbanism and the remaking of contemporary cities*. Taylor & Francis.

IBM. (2021). *Smarter planet*. https://www.ibm.com/ibm/history/ibm100/us/en/icons/smarterplanet/.

Ibrahim, S., & Alkire, S. (2007). Agency and empowerment: A proposal for internationally comparable Indicators. *Oxford Development Studies, 35*(4), 379–403.

Jasanoff, S. (2015). Imagined and invented worlds. In S. Jasanoff & S. H. Kim (Eds.), *Dreamscapes of modernity: Sociotechnical imaginaries and the fabrication of power* (pp. 220–228). University of Chicago Press.

Jasanoff, S. (2015b). Future imperfect: Science, technology and the imaginations of modernity. In S. Jasanoff & S. H. Kim (Eds.), *Dreamscapes of modernity: Sociotechnical imaginaries and the fabrication of power* (pp. 1–33). University of Chicago Press.

Jasanoff, S., & Kim, S. H. (Eds.). (2015). *Dreamscapes of modernity: Sociotechnical imaginaries and the fabrication of power*. Chicago: Chicago University Press.

Jessop, B., Moulaert, F., Hulgård, L., Hamdouch, A., Hulgard, L., & Hamdouch, A. (2013) Social innovation research: A new stage in innovation analysis? In F. Moulaert (Ed.), *The international handbook on social*

innovation: Collective action, social learning and transdisciplinary research (pp. 110–130): Edward Elgar Publishing.
Joly, P.B. (2019) Reimagining innovation. In S. Lechevalier (Ed.), *Innovation beyond technology: Creative economy* (pp. 25–46). Springer.
Kaika, M. (2017). 'Don't call me resilient again!': The new urban agenda as immunology or what happens when communities refuse to be vaccinated with 'smart cities' and indicators. *Environment & Urbanization, 29*(1), 89–102.
Kaika, M., & Swyngedouw, E. (2000). Fetishizing the modern city: The phantasmagoria of urban technological networks. *International Journal of Urban and Regional Research, 24*(1), 120–138.
Keanbirch. (2019, March 26). The in/humanity of technoscience: Environments, imaginaries, and inequalities. *Political Economy of Technoscience*. https://politicaleconomyoftechnoscience.wordpress.com/author/keanbirch/.
Leitheiser, S., & Follmann, A. (2019). *The social innovation–(re)politicisation nexus: Unlocking the political in actually existing smart city campaigns? The case of SmartCity Cologne*. Urban Studies.
Lisbon Declaration. (2018). *Social innovation as a path to a sustainable, resilient and inclusive Europe*. https://www.siceurope.eu/sites/default/files/field/attachment/the_lisbon_social_innovation_declaration15.10_0.pdf.
Maglavera, T., Niavis, H., Moutsinas, G., Passani, A., & De Rosa, S. (2019). *Digital transformation for a better society*. ChiC, European Commission. https://capssi.eu/wp-content/uploads/ChiC_D5.2_Digital_Transformation_for_a_better_society-whitepaper.pdf.
March, H., & Ribera-Fumaz, R. (2014). Smart contradictions: The politics of making Barcelona a self-sufficient city. *European Urban and Regional Studies, 23*(4), 816–830.
Mazzucato, M. (2013). *The entrepreneurial state: Debunking public vs. private sector myths*. Anthem Press.
McFarlane, C., & Söderström, O. (2017). On alternative smart cities: From a technology-intensive to a knowledge-intensive smart urbanism. *City, 21*(3–4), 312–328.
McLaren, D., Agyeman, J., & Gottlieb, R. (Eds.). (2015). *Sharing cities: A case for truly smart and sustainable cities*. MIT Press.
Mechant, P., De Marez, L., Claeys, L., Criel, J., & Verdegem, P. (2011). Crowdsourcing for smart engagement apps in an urban context: An explorative study. *IAMCR Proceedings*.
Misuraca, G., Pasi, G., & Urzi Brancati, C. (2017). *ICT-enabled social innovation: Evidence & prospective*. Publications Office of the European Union.
Moulaert, F., & Van Dyck, B. (2013). Framing social innovation research: A sociology of knowledge perspective. In F. Moulaert, D. MacCallum, A. Mehmood, & A. Hamdouch (Eds.), *International handbook on social innovation*. Edward Elgar.

Mulgan, G. (2007). *Social innovation: What it is, why it matters and how it can be accelerated*. The Young Foundation.

Nicholls, A., Simon, J., & Gabriel, M. (2015). Introduction: Dimensions of social innovation. In *New frontiers in social innovation research* (pp. 1–26). Palgrave Macmillan.

Ozman, M., & Gossart, C. (2019). *Digital social innovation: Exploring an emerging field*. ISIRC 2018 Conference, 5–7 September, Heidelberg. Available at SSRN: https://ssrn.com/abstract=3434363 or https://doi.org/10.2139/ssrn.3434363.

Parayil, G. (2005). *Information capitalism*. Springer SCOOP.IT https://www.scoop.it/.

Patel, M., Sotsky, J., Gourley S., & Houghton, D. (2013). The emergence of civic tech: Investments in a growing field. *Knight Foundation Report on Civic Technology*. https://www.slideshare.net/knightfoundation/knight-civictech/10READING_THE_MAPLandscape_ThemesOpen_Government.

Podder, S., Singh, S. K., Tanguturi, P., Roy, P., Gueret, C., & Dubash, N. (2018). *Tech4Good: Collectively scaling up social transformation*, downscale 18, May 2018, Amsterdam, The Netherlands.

Prize2theFuture. (2021). *About the prize*. https://www.prize2thefuture.org/about/overview.html.

Rodrigo, L., Palacios, M., & Ortiz-Marcos, I. (2019). Digital social innovation: Analysis of the conceptualization process and definition proposal. *Direccion y Organizacion, 67*, 59–66.

Russon, H. (2017). Civic tech for urban collaborative governance. *Political Science & Politics, 50*(3), 744–750.

Schubert, C. (2018). Social innovation: A new instrument for social change? In W. Rammert & A. Windeler (Eds.), *Innovation society today* (pp. 371–391). Springer VS.

Segercrantz, B., & Seeck, H. (2013). The construction of social innovation and- undesirable consequences of innovation—A critical reading of the European Union's social innovation policy. In *EU-SPRI forum conference on management of innovation policies*, 10–12 April 2013, Madrid, Spain.

Seyfang, G., & Haxeltine, A. (2012). Growing grassroots innovations. *Environment and Planning C: Government and Policy, 30*(3), 381–400.

Sharable. (2018). *Sharing cities*. https://www.shareable.net/sharing-cities/.

Smith, J. M., & Tidwell, A. S. D. (2016). The everyday lives of energy transitions: Contested sociotechnical imaginaries in the American West. *Social Studies of Science, 46*(3), 327–350.

Söderström, O., Paasche, T., & Klauser, F. (2014). Smart cities as corporate storytelling. *City, 18*(3), 307–320.

Sovacool, B. K., & Hess, D. J. (2017). Ordering theories: Typologies and conceptual frameworks for sociotechnical change. *Social Studies of Science, 47*(5), 703–750.

Sovacool, B. K., & Ramana, M. V. (2015). Back to the future: Small modular reactors, nuclear fantasies, and symbolic convergence. *Science, Technology, & Human Values, 40*(1), 96–125.

Stokes, M. (2020). *What system factors help DSI to grow and thrive?* DSI4EU, Nesta. https://digitalsocial.eu/blog/145/a-first-run-of-the-dsi-index.

Strand, R., Saltelli, A., Giampietro, M., Rommetveit, K., & Funtowicz, S. (2018). New narratives for innovation. *Journal of Cleaner Production, 197*(2), 1849–1853.

Surowiecki, J. (2004). *The wisdom of crowd.* Doubleday.

Swyngedouw, E. (2005). *Social innovation, governance, and community building—SINGOCOM.* Final Report Month 40, European Union Framework V, Key Action Improving Socio-economic Knowledge Base, Contract No. HPSE-CT2001-00070, Project No. SERD-.

Thompson, M. (2019). Playing with the rules of the game: Social innovation for urban transformation. *International Journal of Urban and Regional Research, 43*(6).

Tidwell, J. H, & Tidwell, A. S. D. (2018) Energy ideals, visions, narratives, and rhetoric: Examining sociotechnical imaginaries theory and methodology in energy research. *Energy Research & Social Science, 2018/05*(39), 103–107.

Törnberg, A. (2018). Combining transition studies and social movement theory: Towards a new research agenda. *Theory and Society, 4*, 381–408.

Tourist Office Ghent. (2019). *Official city guide of Ghent.* https://visit.gent.be/en/good-know/practical-information/publications.

Townsend, A. M. (2013). *Smart cities: Big data, civic hackers, and the quest for a New Utopia.* Norton & Company.

Turkle, S. (2011). *Alone together: Why we expect more from technology and less from Each Other.* Basic Books.

Van Noordt, C., Sobirjonov, A., Thompsoon, J., & Wurst, A. (2018). *E-participation & innovation in local government: A look at urban crowdsourcing in the cities of Antwerp and Ghent.* Research Seminar paper, January 8, KU Leuven.

Vanolo, A. (2013). Smartmentality: The smart city as disciplinary strategy. *Urban Studies, 51*(5), 883–898.

Vienna Declaration. (2011). *Innovating innovation by research—100 years after Schumpeter.* http://www.socialinnovation2011.eu/wp-content/uploads/2011/09/Vienna-Declaration_final_10Nov2011.pdf.

Viola, R. (2020, March 8). *The internet of humans.* European Commission. https://ec.europa.eu/digital-single-market/en/blogposts/internet-humans.

Von Hippel, E. (2005). *Democratising innovation.* MIT Press.

Wittmayer J. M. et al. (2019) Narratives of change: How social innovation initiatives construct societal transformation. *Futures*, *112*.
Zuboff, S. (2019, July 2). It's not that we've failed to rein in Facebook and Google. We've not even tried. *The Guardian*. https://www.theguardian.com/commentisfree/2019/jul/02/facebook-google-data-change-our-behaviour-democracy.

Links

Apps for Ghent. http://www.appsforghent.be/.
Arduino. https://www.arduino.cc/.
Better Reyjkavik. https://reykjavik.is/en/better-reykjavik-0.
BurgerBudget. https://persruimte.stad.gent/159337-burgerbudget-tijd-om-te-stemmen.
Chic - Chicory as a multipurpose crop for dietary fibre and medicinal terpenes. http://chicproject.eu/.
Citizen Foundation. https://citizens.is/.
CitizenLab. https://www.citizenlab.co/.
City of People 2019. http://citiesofpeople.com/.
Citying platform. http://www.citying.ethz.ch/.
Commons Network. https://www.commonsnetwork.org/.
Commons Transition Plan. https://www.commonstransition.org/commons-transition-plan-city-ghent/
Crowdsourcing Gent. https://crowdfunding.gent/nl/.
DeCommunity. http://decommunity.be/.
Dienst Beleidsparticipatie. https://stad.gent/over-gent-en-het-stadsbestuur/stadsbestuur/organisatiestructuur/bedrijfsvoering/dienst-beleidsparticipatie.
EuroCities. https://eurocities.eu/.
Fixmystreet. https://www.fixmystreet.com/.
GentSintPieters. https://www.projectgentsintpieters.be/%20%20https:/samenlevingsopbouwgent.be.
Ghent Living Lab. https://stad.gent/smartcity/ghent-living-lab.
Ghent Smart City. https://stad.gent/smartcity-en.
Guifi. https://guifi.net/en.
Ideeen voor MMM. https://ookmijn.stad.gent/mmm-ideeen/ideeen.
Impossible living. http://blog.impossibleliving.com/.
Lleialtat. https://www.lleialtat.cat/.
Manouvre. http://www.manoeuvre.org/.
MK:Smart. http://www.mksmart.org/.
New Economic Foundation. https://neweconomics.org/.
Open Ministry. https://www.oministry.com/.
Otwarta Warszawa. https://www.otwartawarszawa.pl/.

P2P Foundation. https://p2pfoundation.net/.
Paris Idée. https://idee.paris.fr/.
Prize2theFuture. http://www.prize2thefuture.org/.
PublicLab. https://publiclab.org/.
Refill. https://urbact.eu/Refill.
Smart Citizen. https://smartcitizen.me/.
Speak. https://www.speak.social/en/SpeakFE/online-language-groups/.
Stads Academie. http://destadsacademie.be/category/sessie/.
Timelab. https://timelab.org/.

CHAPTER 5

Reproduction: Digital Social Innovation in Urban Governance

Abstract The present chapter explores how Digital Social Innovation (DSI) initiatives contribute to the processes that reproduce society by feeding new forms of urban governance. Notably, it analyses how these initiatives are mobilised in governance projects and produce technology-enhanced social configurations. In this way, they can reinforce, rethink, or contest the neoliberal urban agenda by innovating both the coded sets of procedures and the un-coded routines that make a city work. Therefore, the chapter explains how DSI initiatives are used to materialise political aims through digital artefacts and procedures.

Keywords Digital social innovation · Social reproduction · Urban governance · Neoliberal governance · Governance beyond-the-State · Technological sovereignty

5.1 Introduction

In the previous chapters we discussed DSI as a social technology, whose definition is still subject to interpretative flexibility and whose practical realisations are inspired and characterised by the different intents of the community of innovators. Building upon Misuraca et al.'s classification of SI (2017), three alternative approaches to DSI have been identified.

C. Certomà, *Digital Social Innovation*,
https://doi.org/10.1007/978-3-030-80451-0_5

A first category includes DSI initiatives characterised by the strong presence of the business sector, which functionally contributes towards the materialisation of the neoliberal rationality of digital capitalism. The role of local public institutions or civil society organisations become more prominent, we can identify a second reformist category of DSI initiatives, that includes a broad variety of nuanced approaches whose common trait is the willingness to promote a harmonic orchestration of public and private interests in pursuing social goals a last category includes revolutionary initiatives intended to subvert the status quo. Such a distinction is pivotal in the analysis of the multiple ways DSI processes contribute to social reproduction through urban governance. Overall, the analysis of imaginaries supporting DSI makes it evident that it cannot be described as a monolithic phenomenon but rather as a polyvocal set of practices performed in support of the reproduction of a distinct reading of a city identity and plans for urban development. As a matter of fact, alternative DSI imaginaries and related practices are not elaborated in isolation, but rather in constant dialogue with diverse urban imaginaries, emerging from encounters between digital technologies and social innovation aims. We already pointed out that imaginaries are capable of being performed and therefore flow from imagination to plans, conception and eventually realisation; and governance processes make this transformation possible because they turn imaginaries into reality. The augmented city generated by the digital revolution is prolific in terms of narratives, discourses, metaphors and visions. The imaginaries associated with DSI initiatives are mobilised in socio-technological projects and produce technology-enhanced social configurations (while also illustrating DSI as a social technology).

In the context of a growing diffusion of neoliberal models of urban development, infiltrating and changing urban governance processes, DSI initiatives are unavoidably entangled with reproducing a neoliberal society or in contesting it, by innovating both the coded set of procedures and the un-coded routines that make a city work.

By adopting a critical geographical perspective, we can say that DSI initiatives are used as a means for social reproduction and they do so by materialising political aims into digital artefacts (e.g. the open-access model, the federated internet project and similar) and by shaping the augmented urban space through the social construction of technologies (Glimell, 2001). For instance, blockchain, crowdsourcing, e-democracy or e-learning protocols are frequently adopted into DSI projects to address emerging socio-political challenges. At the same time, these also create new spaces of interaction, with their own rules and constraints, enabling material and informational infrastructures, new sites of command and control, and cracks from where the system can be hacked. This condition unavoidably affects traditional forms and function(ing) of urban governance (Chesbrough, 2003; Von Hippel, 2005). The collective agency of digital social innovators infiltrates and modifies social organisations and (government, research and business) institutions; it generates unedited social configurations and previously unthinkable political, economic and environmental scenarios.

The present chapter explores how DSI initiatives contribute to processes that reproduce society by feeding new forms of urban governance. These are characterised by the intense and fluid relationships between public, private and hybrid actors, and contribute to the emergence and legitimation of the global digital governmentality.

5.2 Governance as a Mode for Social Reproduction

Governance is not limited to the sphere of government. Rather, it can be defined as the ensemble of all those formal and informal processes undertaken by governing bodies and administrations at different geographical scales (global, regional, national, local), market organisations, civil society associations and individuals in order to influence the fate of an issue of public relevance. By adopting multiple tools (including norms, behavioural patterns, media or economic models) and various modes of social interaction (including negotiation, conflict or dialogue), governance processes produce non-occasional forms of social ordering, which allows involved actors to develop contingent plans and coordinate following agreed-upon intentions (Hofmann et al., 2016).

From the '70s, western democracies have been characterised by the progressive transformation of top-down public government processes and the privatisation of public services (Hodge, 2000). Two decades later, a generalised condition was achieved, in which most traditional government functions were accomplished through governance processes (Rhodes, 2007). This shift from government to governance brought about crucial changes in the perception and realisation of social re-productive processes. The increase of economic and operational capacity of private actors and, conversely, the decrease of authority and resources available to public institutions, determined a progressive devolution of public administration to horizontal networks of public-private associations from the '80s onward. Nevertheless, Thomas Lemke noted that "What we observe today is not a diminishment or reduction of state sovereignty and planning capacities, but a displacement from formal to informal techniques of government and the appearance of new actors on the scene of government (e.g. NGOs), that indicate fundamental transformations in statehood and a new relation between state and civil society actors" (Lemke, 2002, p. 50). By gathering heterogeneous actors around a common matter of concern, these new "techniques of government" make actors traditionally excluded from the business of politics and administration able to impact decision-making and public management processes, which were once the prerogative of public actors.

The capacity to influence complex social systems in the direction of their interests (Hamel, 2002) and to reach satisfactory decisions *via* free negotiation not burdened with legal or ethical constraints, is a key trait of non-state actors, that allows them to deal with a broad range of problems (including those lying outside of the sphere of traditional government).

A networked organisation requires interdependence to ensure that the agency of independent actors is effective. At the same time, the interaction needs to be regulated with ad hoc agreements because traditional regulatory frameworks are often inadequate in an increasingly privatisatised context. The consolidation of the network model in governance processes, which often completely replaced traditional government ones, brought along unexpected consequences associated with the proliferation of flows of (economic, financial, social, cultural, technological and symbolic) capital from one node to another of the network. For instance, it determined the deregulation of many public sectors and the fragmentation and decentralisation of competences and tasks, with the concurrent proliferation of command and control centres.

The transition from traditional, public institution-driven decision-making processes towards those that involve multiple actors has been discussed in participatory governance literature, in which a vast range of possible interaction modes have been described (Brabham, 2009). The diffusion of bottom-up governance processes has been welcomed as an opportunity for civil society associations and grassroots social movements to be officially involved in policy processes and to directly negotiate with political decision-makers to ultimately advance their claims (Acuto & Rayner, 2016; Portney & Berry, 2013; Seyfang & Haxeltine, 2012).

However, the involvement of a wide range of non-institutional actors (including small business and big corporate, intermediate bodies, international institutions, research centres and academia, further than CSOs, NGOs and citizens informal associations) generated a magmatic universe of institutional, quasi-institutional or informal networks in which the attribution of roles, responsibilities and benefits is not always driven by socially emancipatory values. It is, in fact, not uncommon to finds out that these heterogeneous governance configurations do not foster greater democratisation, sustainability, justice or, in general, socially progressive objectives. Put simply, governance processes are ways in which society reproduces itself—and therefore include reformist, revolutionary and also conservative practices that mobilise diverse forms of capital.

Despite being intended originally as a variation of the economic reproduction process, social reproduction processes are not limited to the economic sphere but refer to multiple aspects of the private ones, such as the procreation, care, education, religion and culture (Cammack, 2020). As Pierre Bourdieu (1984) noted, different types of capital contribute to social reproduction, including the economic, cultural (for instance competencies or skills), human (such as education level and professional qualifications) and social (notably the constitution of social networks and the creation of trust). Together with re reproducing the material existence of society, these also transmit from generation to generation, the symbolic capital (i.e. honour or reputation), which allows a person to exercise authority and even violence on others endowed with a lower degree of this form of capital; and in consequence consolidate social stratification (Bourdieu, 1984). Today a further form of capital compliments existing ones and contributes to the process of social reproduction, this is technological capital (Ignatow & Robinson, 2017). In management studies, technological capital is defined as the combination of "the tangible component, including the active part of the firm's tangible fixed assets, and the intangible component, comprising intangible assets related to products manufacturing and production management" (Grigoriev et al., 2014, p. 56). Other social research disciplines devoted comparatively less attention to technological capital and often used the concept to refer to an individual's ability to take advantage from technological readiness (including knowledge or access possibilities) to overcome the digital divide (Carlson & Isaacs, 2018). Nevertheless, this offers important insights for the analyses of society's entanglement with digital technologies and the broad set of related practices.

The recent shift from government to governance processes obliged us to reconsider how the material and symbolic reproduction of space and society occurs through complex events in which digital technological capital is ubiquitous. Manuel Castells extensively described the contemporary public sphere as a space of debate increasingly constructed around global communication networks in the space of flows (Castells, 2008). DSI initiatives are integrally constitutive of these flows that feed contemporary cities; and work as social technologies to influence the allocation of roles, benefits and responsibilities in governance networks, by raising questions on their actually transformative potential. Whether DSI-infiltrated governance processes really are accessible and empowering in comparison to government-led processes—and who is benefitting from them, is an open issue. This is particularly relevant now that the solidity of public institutions is undermined by the concentration of different forms of capitals in the hands of those whose agency is driven by private interests.

Similar questions on DSI intervention in governance processes emerge from more general critical considerations on the tensions and contradictions characterising current neoliberal governance. These have been captured by critical scholars' reflections on the neoliberal city.

5.3 Tensions and Contradictions in the Neoliberal Governance

The shift from government to governance, which characterises contemporary forms of public life, is closely linked with the consolidation of new technologies of government and with the transformation of traditional parameters of democracy. The idea of governance entails the possibility to improve government processes thanks to the collaboration between the public and private sectors (Stoker, 1998). The availability of new non-State resources promises to make the government more efficient and effective (Pierre, 2000). Thanks to interactions among supposedly equal participants with regular exchanges and wide access to decision-making for representatives of different social categories, governance processes are often described as horizontal, networked and based on interactive relations between independent and interdependent actors. The resulting horizontal networks are polycentric and characterised by distributed and trans-scalar forms of power (Hajer & Wagenaar, 2003). For instance, these networks gather urban committees, CEOs of multinational companies, informal citizens associations, transnational organisations—such as the *World Trade Organisation* or the *International Monetary Fund*, and representatives of public institutions.

The new aggregates made from heterogeneous actors govern from outside and beyond the State (Hajer, 2003; Mitchell, 2002; Rose & Miller, 1992). Geographer Erik Swyngedouw proposes the term "governance-beyond-the-state" to signify "the emergence, proliferation and active encouragement (by the state and international bodies like the European Union or the World Bank) of institutional arrangements of 'governing' which give a much greater role in policy-making, administration and implementation to private economic actors on the one hand and to parts of civil society on the other, in self-managing what until recently was provided or organised by the national or local state" (Swyngedouw, 2005a, p. 1992). This governance-beyond-the-state is characterised by increasing privatisation and deregulation of State functions, and their externalisation through upscaling processes (delegating tasks to higher levels of governance, such as international organisations or macro-regional institutions) and down-scaling processes (involving local, notably urban, administrations in the process of vertical decentralisation). Such restructuration is embedded within the neoliberal political project "that proposes that human well-being can best be advanced by liberating individual entrepreneurial freedoms and skills within an institutional framework characterised by strong private property rights, free markets, and free trade" (Harvey, 2005, p. 2).

However, critical thinkers signalled that this massive change generated a democratic deficit that was particularly evident in social tensions resulting from the progressive erosion of the public sphere (and space) produced by market forces.

The creation of network formations governing society instead of the government do not automatically impede the formation of hierarchies or power asymmetries, neither does it impede the enactment of exclusionary practices. Horizontal network rules can be produced un-democratically and even without agreement from concerned actors (Akkerman et al., 2004; Dryzek, 2000). In time, it became apparent that the democratisation induced by the reversal of top-down into bottom-up processes particularly benefitted already powerful actors (i.e. economic, political and cultural leading elites), while leaving the majority of marginalised people with the burden of competing to have their voice heard in an increasingly crowded, ephemeral and noisy public sphere. As a consequence, "the non-normative and socially innovative models of governance as non-hierarchical, networked and (selectively) inclusive forms of governmentality, cannot be sustained uncritically" (Swyngedouw, 2005a, p. 2002).

In a governance-beyond-the-State context, the ambiguity deriving from the absence of clear channels of representation and accountability makes it even more difficult for civil society to engage in public debate on equal terms with more influential actors. The matters of concern include issues of entitlement, i.e. who is authorised to participate in public decision-making; legitimate representation, because "new groups of participants enter the frame of governance or reinforce their power position, while others become or remain excluded" (Swyngedouw, 2005a, p. 2001); and accountability of non-state or informal actors for the consequences of their operating in the public sphere. Despite the concerns over advancing the agency of business companies, international organisations and financial centres, the governance model has substituted government almost everywhere with the blessing of these very movements (see Castells, 1996). The informalisation of traditional government processes allowed new participants the possibility to infiltrate the ganglia of power. However, it was the very same crack in the system that allowed other, incredibly more powerful, market institutions to have their say in the public decision-making process. Swyngedouw (2005b) describes governance beyond-the-state as a trojan horse that, under the appearance of horizontal configurations empowering civil society in the face of an oppressive state, actually allowed and consolidated the "market" as the principal institutional form. Therefore, while the shift from government to governance opened up the public sphere and its decision-making processes beyond traditional public institution it is also evident now that, superseding the wishful ideal of equality, democratisation and participation, deregulation has also "enhanced the power of groups associated with the drive towards marketisation and has diminished the participatory status of groups associated with social-democratic or anti-privatisation strategies" (Swyngedouw, 2005b, p. 2003). While initially welcomed by non-governmental associations, citizens' movements, and dissidents as a way for breaking the ubiquitous power and the oppressive control of the State, governance processes progressively turned into powerful social technologies for the erosion of the public sphere, replaced by the occupation of all spheres of social life by the market and the affirmation of neoliberal rationality in political agenda. This context of neoliberal governance, with its contradictions, is the condition in which different forms of DSI processes proliferate and are implemented for diverse ends.

It is evident that DSI processes significantly rely on the possibilities offered by networked governance and contribute towards its very establishment as the best way of conducting public life (Moulaert et al., 2005). In fact, being a special form of SI, DSI relies on the possibility of gathering individual and collective social actors, coming from diverse contexts, to address the "intractable nature of social challenges such as poverty, inequality and environmental degradation" (Tracey & Stott, 2017, p. 52) and for promoting ideas and practices that underpin social change by combining public and private agency. It needs to be cautious, however, about what form of governance DSI initiatives contribute at enacting. Critical scholars claim that DSI contains "both utopian and dystopian possibilities for new forms of sovereignty" (Thompson, 2019, p. 1178) and raised doubts about the real potentialities of DSI to foster wide participation and to serve the public good. Certain traits of DSI make it particularly apt for the reproduction of a neoliberal form of governance: DSI aspirations for transforming the meaning and practices of participation (Baccarne et al., 2014); the difficulty to govern DSI in the newly emergent public-private governance regime; and the new power geometries that digital innovation brings about (Bendiek et al., 2019; Parayil, 2005). As we noted in Chapter 3, functionalist DSI practices easily associate with the imaginary of the hyperconnected-city and fit within the neoliberal political agenda. However, contradictions emerging in the governance beyond-the-state model opens the way for the other forms of DSI (notably the reformist or revolutionary ones) that equip governance processes in service of alternative, anti-neoliberal aims (i.e. against the marketisation, privatisation, deregulation, individualisation of social relationships and public services). Therefore, while in its functionalist form it can be regarded as an expression of digital capitalism, reformist and revolutionary DSI initiatives are also promising in terms of openness and inclusiveness. In the following sections we consider how DSI is a constitutive element of urban governance and works as a social technology for society construction and how different forms of DSI contribute to social reproduction processes within or against the neoliberal city development.

5.4 Urban Governance in Time of Ubiquitous Digital Technologies

As we already noted (see Chapter 2) the city serves as the elective site for experimenting with innovative governance processes (Brenner & Theodore, 2005). This is due to specific characteristics that differentiate urban governance from higher scales of governance. For instance, intermediaries are more easily cut-out in the urban rather than in the national or supranational government; being physically proximate and daily experienced, the matters of concern debated in the urban public sphere attract the wide interest of local dwellers; and the existence of local-based norms and agreements that regulate civic interaction in the political, economic, social or cultural sphere allow a more direct intervention with citizens in decision-making. Moreover, the close relationship between citizens and institutions makes it easier to tailor DSI initiatives on local needs and subsequently to upscale the technologies these build upon (Barcelona Activa, 2018, p. 11). On this point it is important to signal that dedicated investments have been recently made in many European cities to improve their technological readiness. The opening of offices dedicated to innovation and technology (e.g. *LOTI* innovation office in London city administration or the equivalent, called *6Aika*, in Finland) or data analytics (e.g. provided by Amsterdam or Barcelona city administration), or the definition and implementation of codes of digital practice and policy directives (e.g. *Infoshare* in Helsinki) are all examples of commitment in European cities to digital governance transformation. However, the key to effective and successful DSI initiatives is not in technological readiness and innovativeness per se, but rather in the collective efforts of networked people to make social technology work smoothly and adequately for social transformations. As the final report of the TEPSIE project clearly pointed out with reference to the nature of the digital technologies in social innovation processes: "Given the often small scale, low e-skills and limited resources of many social innovation initiatives, it is often not leading-edge but existing off-the-shelf technologies that are used. Technology is also, of course, not always involved or needed in social innovation, and many other factors are often equally or more important, such as organisational and financial conditions, available skill sets, the prevailing institutional and social landscape, political issues and priorities, as well as culture, norms and values. Thus, technology must always be seen in its intimate mutual relationship with the actual world of people, things and places, and not least the digital skills which social innovators, and sometimes their target groups as well, have or do not have" (Millard & Carpenter, 2014).

Every DSI initiative can be regarded as the emergent effect of a network of networks. The emergence of a DSI process is produced by the gathering of a collective that provides a core innovative technology. In working, testing, improving such technology (be it a tool—such as an application, a platform or a sensor; or a process—such as a participatory budget or a crowdsourcing algorithm) the collective continuously reframes the problems it is intended to solve. This produces new knowledge and identifies unexpected implications, which in turn stimulate the creation of further linkages with other actors, by making the boundaries of the system extremely porous and fuzzy. The agency of DSI collectives generates further networks that may emerge, for instance, from the relationships among similar initiatives (e.g. all the collaborative urbanism digital platforms); or initiatives financed by the same founder or promoted by the same promoter (e.g. all the DSI initiative funded by the EC); or cognate initiatives happening in the same city or region, aiming at the same goal or even adopting a similar set of technologies.

By taking advantage of the key properties of digital networks (including decentralised access, distributed outcomes, simultaneity and interconnectivity), DSI collectives often connect and cross-fertilise each other and cover a broad variety of intervention fields. They often operate in the grey space between public and private (Brenner & Theodore, 2005) and give raise to unedited social, economic and political configurations affecting traditional forms and functioning of governance (Chesbrough, 2003). While the entrance of disparate new actors in the urban governance sphere has been extensively explored already (notably, Castells, 2008; Sassen, 2004), digital social innovators adopt novel forms of agency that conjugate the top-down with bottom-up approaches into mixed patterns generally described in participatory and collaborative approaches. Therefore, DSI initiatives, in general, exploit the prerogative of the networked governance model and equip it with new technological tools and web-based problem-solving processes (e.g. smartphone software, blogs, wikis, social bookmarking applications, social networks, peer-to-peer software, social mapping, geo-referenced social networking, co-designing, cloud processing, etc.). The resulting form of governance is characterised by continuous restructuration of power and responsibility relationships between multiple public and private actors operating beyond (and behind) the local or national governments.

5.5 Digital Social Innovation in Urban Governance

As described in Chapter 2, many DSI processes are fully functional to the creation and reproduction of neoliberal governance regimes through (often non-horizontal) collaborations of government and business and in so doing, imbue market-based values into urban ecosystems (Carayannis et al., 2012; Lindberg, 2011). These kinds of DSI processes, presented as accelerators of urban development, fully integrate with the logic of public–private partnerships in which big commercial companies take part or even lead public decision-making processes, without questioning their political underpinning and socio-economic consequences. Even though they evoke participation, connectivity and democracy ideals, they contribute to the depoliticisation of public concerns through the deemed neutrality of the technology-led public management systems they promote.

For instance, *Citizens Lab* is a fast-growing private company based in Brussels. Since 2015 it has leveraged public-private agreements to advance civic tech solutions supporting city councils' operations. It provides services for public administrations interested in realising participatory planning, budgeting or voting processes. These allow citizens' to intervene in public discussion and decision-making through all phases of public government, from planning can decision-making. Different toolkits are offered together with technical assistance for creating surveys, online workshops, options analysis and similar. Public administrations can select and buy the solution that fits with their needs best. The support packages are proposed as politically neutral products. This is one of the main traits of DSI initiatives which nicely integrates into the reproduction of neoliberal governance processes; they present themselves as the norm, the obvious choice, and therefore need no justification because of their adherence with the mainstream. A clear example of how functionalist DSI with the seemingly progressive character of technical innovation actually supports powerful economic elites, comes from the US. The *Fab Academy* is a large digital fabrication campus that distributes a model of education that globally attracts students on the promise that everybody can easily learn prototyping, planning and executing a new project. Started in

2010 it is supported today by big companies such as *Orange* or *Qatar Development Bank*, and runs in more than 70 fab-labs paid by substantial individual fees, offering learning opportunities to students while creating job opportunities for the teachers. Like the *Fab Academy*, many functionalist DSI projects operate between small enterprises and social services offered in the market. It is difficult in these cases to clearly separate social benefits from private interest; the private interest of many may lead to wide social benefits. Yet, when private interest becomes the priority over collective benefit, then the market-based rationale of these initiatives becomes apparent.

The traits that exactly differentiate initiatives resides in the underlying ideals and worldviews (e.g. market-based, social-democratic, technocratic, communitarian), the approach (i.e. supportive or contrastive of existing institutions or the social organisation), the adopted means and tools (e.g. use of proprietary or non-proprietary software) and the actors gathered (selected, for instance, on their compliance with ethical principles or not). In fact, it might well happen that the distinction between different or even opposite forms of DSI become fuzzy in their practical realisation because they all commit to addressing issues that negatively impact urban life where the same problems (e.g. the lack of child-care facilities) can be addressed through different ways. How in practice DSI initiatives are performed, what technologies they adopt and who is funding them, makes a difference in their positioning within or outside neoliberal governance.

Differently from functionalist projects, reformist DSI initiatives sit in a grey space between the neoliberal mainstream and the revolutionary intents of cyber activism. These often revolve around small adjustments to existing governance processes. For instance, the *SPEAK* project (see Chapter 1), started in 2012 by the Portuguese non-profit organisation *Associação Fazer Avançar*, aims to promote cultural diversity and equality by organising language learning groups for migrants and refugees, which integrate online and offline talks under the payment of a symbolic fee. The platform today has 36,000 users, with 47 languages shared in 28 hosting cities. The overall aim is to facilitate social integration challenges by leveraging on people's ability to change the world they live in *via* mutual help and networking. Gathering a broad partnership of over 30 supporting partners (including migrant associations, students exchange associations, hostels, local libraries and foundations) *SPEAK* helps non-native speakers learn the local language (and also to provide lessons for others). The initiative is economically self-sustained and designed to be reproduced (under franchising licence) in diverse cities, but it relies on proprietary technology developed by the private company, *Outsystem*. Another initiative which aims at reforming existing urban governance routines is *Transparent Krakow*, an open data programme aimed at regularly monitoring public authorities operating in Krakow. Advanced by the civil society organisation *Stanczyk Foundation*, it gathers information on issues that are important to local residents, such as session records of the City Council, local committees and district councils. It casts upon the biggest open data repository in Poland, *mojepanstwo.pl*, an open-source portal created by *ePaństwo Foundation*—an independent NGO. The difficulty to get economic support for conducting monitoring campaigns which are important to democracy is the price of independence. This suggests that we need to analyse the technological components of DSI initiative to understand their positioning. Considering these components tells us about who is funding the project; who is involved; who economically gains from it; who is expected to get a benefit in terms of services; and whether the project is aimed at making existing processes more efficient or effective or to radically change them?

Digital activists, whose agency often generates what we have defined as revolutionary DSI initiatives, already pointed out the need for a critical engagement with digital technologies and their socio-political impacts in the augmented urban space. For instance, a large movement is assembling around the issue of "technological sovereignty" against the smart city rhetoric and the abuses of platform companies (like *Airbnb*, *Uber*, and *Deliveroo*) and against monopolies on digital infrastructures. The movement of community-owned network infrastructures promotes crowdsourcing to extend internet connectivity *via* non-proprietary hardware and software. The *Athens Wireless Metropolitan Network* founded in 2002 is an example of software routers and dynamic routing infrastructure. A similar attempt has been advanced in Ghent, where the *Small Tech Foundation* (former *Ind.ie*), a not-for-profit association that develops small tech "designed to increase human welfare, not corporate profits" designed a federated personal web sites system for the city.

TextBox 5.1. Technological Sovereignty the Catalan Way

In Barcelona the claim for digital sovereignty has been embraced by the regional government since 2011 (Barcelona Digital City, 2020). In practical terms the local administration envisages "a broad, de-centralised network of cooperatives, associations, and community initiatives experimenting with alternative practices of locally rooted, open-source digital development" (Lynch, 2019, p. 660). An example of relocating innovation in society is offered by the *SocialHack Nou Barris*, the first hackathon held in a deprived area of Barcelona, Nous Barris. Bringing technologies and related initiatives where there are none or where people lack any form of civic engagement has been a pivotal attempt at seeking out solutions to the district's most pressing social challenges.

The engaged scholar Casey Lynch reports two cases in his research blog. The first case is provided by the platform *Katuma*, which links data sovereignty with the local food economy: "Katuma [is] a platform cooperative working to promote local food economies. The local programming cooperative, *Coopdevs*, works to build and maintain the digital platform infrastructure, based on the *Open Food Network* platform, that enables the coordination of a local, organic food economy. As a cooperative, *Katuma* is owned and controlled collectively by local producers and consumers. The use of an open-source platform maintained by another local cooperative helps assure that *Katuma* is able to maintain control over their own data and make collective decisions about how the platform operates. Technological sovereignty facilitates food sovereignty" (Lynch, 2020). The second case is the famous Guifi.net initiative, an energy and digital sovereignty programme in Catalonia: "the community wireless network, Guifi.net [...] has begun experimenting with open-source DIY home automation and IoT sensing technologies, focussing initially on home energy monitors in the *CanGuifi* project. The initiative is meant to design low-cost technologies that allow residents to effectively monitor their energy usage without ceding that data to private energy companies, who often exploit that data through new pricing mechanisms or other programmes. The project has the potential to create new partnerships with a local movement around energy sovereignty, at least partially focused around the renewable energy cooperative, *Som Energia*" (Lynch, 2020). In this case the DIY approach to digital technologies allows people's control on their energy production, distribution and use.

A similar example is provided by the *Lleialtec* project, whose intent is to raise "people's awareness about the importance of technological sovereignty by implementing infrastructures that enable civic management facilities to take control of their data and networks" (Lleialtec, 2020). In 2017, an opinion movement among Catalan citizens called for digital empowerment and community action to independently manage and develop digital communication and data storage. Building upon Guifi.net network infrastructure, two free and open-source software digital tools were produced: *Àgora*—a forum that serves as a permanent online assembly, and *Núvol*—a cloud where data is stored.

Moreover, the *Decidim* project creates an open-source digital platform for citizen participation, in which citizens can make proposals and contribute to the development of municipal initiatives. Originally created by digital activism, the platform is now used by municipal governments and organisations across Europe. Also, *Ateneus de Fabricacio´* and the *Barcelona en Comu´* have promoted "changes to municipal practices with an emphasis on free software, open data, transparency, and citizen participation" and the administration migrated from proprietary software package and reviewed its municipal contracting guidelines to support local cooperatives and firms based on open-source technology and social values. As Lynch correctly signals (2019), rejecting neoliberal, hyperconnected and smart city programmes does not imply a rejection of digital innovation for social development, which can become the site for imagining alternative logics and building alternative practices.

Whilst it is obvious that digital technologies adopted in revolutionary DSI processes are often small scale and low budget, the path they are tracing is nevertheless crucial to understanding future developments. They constitute seminal attempts at addressing the very core of social reproduction processes in the digital age because by questioning and remaking the hard and soft infrastructures of ICTs, and undermine the material base of neoliberal governance routines. It is not surprising, than, that revolutionary DSI initiatives become an object of interest in the *Future Internet Research and Experimentation* strategy by the EC (EC, 2020). As a consequence, they have been, at least partially, integrated in reformist strategies of some EU projects. For instance, the *Smart Citizen* project created a low-cost set of tools to collect, transmit and elaborate data on environmental quality. With over 2400 kits distributed in European cities it adopted *Arduino* technology, an open-source electronic prototyping platform enabling users to create interactive electronic objects. The collected data are transmitted by wi-fi to a server to be analysed and discussed by citizens themselves *via* the open-source *Smart Citizen* Platform through participatory projects. Universities and city administrations can use the data for free.

Academic research is also exploring revolutionary DSI practices. Amongst the others, the *Grassroot Innovation* project at the University of Sussex and East Anglia explains that hackerspaces and fab-labs are the most prolific contexts for critical social innovation to emerge (Smith et al., 2014). Researchers are interested in analysing the working of "community workshops where people can access versatile, digitally-controlled manufacturing technologies, open-source designs and training, and like-minded people working on collaborative projects in a wider 'maker' social movement. Community workshops connect online through social media and physically in meetups and conferences. Workshops and their networks enable people to work together in practical, hands-on projects" (Grassroots Innovation, 2020). More interestingly, they are investigating how, while creating objects, people engage in the critical design of entire processes that "can reconfigure, relocate, and recalibrate innovation capabilities in society" (Grassroot Innovation, 2020).

In general, the integration of DSI projects in urban governance, depending on their very nature, can contain both utopian and dystopian possibilities. In the digital age, technological artefacts are fully constitutive of our society where all technologies are social technologies (Latour, 2005), so that the techne and episteme are associated in a single techno-scientific rationality. ICTs, social webs and the IoT are associated with specific techno-scientific rationality where physical and social space is mediated and produced in the digital space and *via* digital means, that "on the one hand enforce new forms of (self)disciplining, and on the other enact new forms of control" (Ash et al., 2018, p. 31).

DSI initiatives are able to operate towards the social reproduction of different organisational forms, material settings and ideal purposes, depending on the aims of the promoters. As Thompson insightfully noted: "For instance, Blockchain technologies—underpinning cryptocurrencies like Bitcoin—have the potential to 'enable modes of large-scale collective action that bear very little resemblance to government as we've known it', by circumventing the hegemonic medium of money and its ultimate arbiter, state power, through a radically decentralised, accountable and transparent digital ledger system (Greenfield, 2017, p. 161).

New production processes around digital fabrication, enabled by the technological innovation of 3D printing and embodied socially in maker spaces (Greenfield, 2017), have the potential to short circuit capitalism by making the production of commodities approach zero marginal cost, thereby undermining the commodity form (Mason, 2015; Srnicek & Williams, 2015). Makerspaces provide tentative infrastructures for more collaborative production (van Holm, 2017) and 'open innovation', which transcends closed, vertical control by powerful corporations (Chalmers, 2012)" (Thompson, 2019, p. 1178).

However, this progressive and emancipatory capacity is only exerted when public interest is prioritised over the private one. Therefore, we always need to be extremely cautious when considering the integration of DSI initiatives into urban governance processes, because their performance is never a neutral act. It is through the combination of the material performance and their embedded values that these can advance or contrast the reproduction of neoliberal governmentality. Critical approaches and revolutionary initiatives prefigure the possibility of re-politicising the debate on DSI by taking into consideration the creative potential of dissensual digital practices. These represent the object of the next chapter.

5.6 Towards an Understanding of Digital Social Innovation as a Political Action

In the previous chapters, we explored the world of DSI and established it as a multifaced phenomenon whose forms, aims and effects largely depend upon the visions, competences, contexts of performance and objectives of the involved communities. As we will see in the next chapter, a critical reading of DSI opens up important considerations on its socio-political implications.

The mainstream imaginary and understanding of DSI presents a set of processes smoothly integrated with neoliberal governmentality and is functional to the reproduction of the hyperconnected-city. The functionalist perspective (Dacin et al., 2011) emerged as "predominantly a practice-led field in which definitions and meanings have emerged through people doing things in new ways rather than reflecting on them" (Mulgan, 2007, p. 10). This approach has devoted significantly larger attention to the process through which social innovation is realised

(Millard & Carpenter, 2014) rather than to the socio-political assumptions triggering it. Nevertheless, we discussed alternative and potentially revolutionary DSI practices that can generate a re-politicisation effect. As Lynch correctly noted: "The de-politicisation of key aspects of urban life through the implementation of 'smart city' models is just the latest in a long succession [...] focused on making cities 'competitive, global, secure, and sustainable' (Davidson & Iveson, 2015, p. 544). This constructed consensus as to what constitutes good urban governance allows for the rise of technocracy—as experts are brought in to implement global 'best practices' and the space of democratic debate is continually constrained" (Lynch, 2019, pp. 665–666). Neoliberal governmentality replaces democratic debate (and dissensus) with data-driven government models. Nevertheless, while being blamed for bringing about grievous consequences in terms of both physical and metaphorical erosion of the public space and authority under the pressures of the global capitalism, it also invites social activism to manifest its disagreement towards the current state of things.

The unprecedented fluidity of the public sphere, while advantaging the new digital elites, also allows new actors from below to infiltrate the ganglia of global and local governance, and to originate a re-politicisation moment (Diaz-Parra et al., 2015). Notably, in the governance-beyond-the-state model, politics displaced from traditional loci of democratic debate and substituted by massive economic interests, is re-emerging in the digital domain. This is principally due to the commitment of social and community enterprises, formal or informal associations, CSOs, NGOs or even public institutions, whose efforts are primarily aimed at the "transformation of institutions, overthrowing oppressive structures with power, collective agency to address non-satisfied needs, the building of empowering social relations from the bottom-up" (Moulart & Van Dyck, 2013).

DSI initiatives offer at once new possibilities and new threats. The networked structure of the city is today amplified by the quasi-continuous usage of personal digital devices that turn it into a hypertextual space endowed with the versatile structure of knowledge and power (Andrejevic, 2016).

The apparent communication and exchange freedom facilitated and distributed by peer-to-peer networks represent, in Foucauldian terms, a technique of governance that produces distinctive power relationships characterised by decentralised control (Galloway, 2004). This fluid form of power corresponds to Swyngedouw's governance beyond-the-state described earlier. It brings about "the restructuring and re-organisation of governing technologies [...], experimental arrangement of new discourses of negotiation, entrepreneurial self-empowerment strategies, and social organisation. In this type of culture, power no longer comes from a central source, but is generated by the ongoing shape-shifting of its members, who it involves in procedural methods of defining and valuation practices". (Reichert, 2017, p. 38).

In re-organising governance technologies, digital collectives are deemed the torchbearers of the ongoing social revolution (Shirky 2008), hailed as the caretaker of collective intelligence (Surowiecki 2004). These collectives we described as digital social innovators provide a versatile modus operandi in which consensual opinion formation is not reached through traditional democratic processes but is embedded in the structured design of social software. This means that the power of the many, produced by digital networking without a central mobilising authority, becomes imperative (Reichert, 2017). As Reicher explains, innovators' collective agency is increasingly permeating the governance of contemporary cities and "appear as poorly-defined systems characterized by weakly-structured volumes of data and a similarly fuzzy logic" (Reichert, 2017, p. 41). The condition they generate has been named as "algorithmic governance" (Müller-Birn et al., 2013), a mode "of social ordering that relies on coordination between actors, [that] incorporates particularly complex computer-based epistemic procedures" (Katzenbach & Ulbricht, 2019, p. 3). Similar governance structures are perfectly fitting with the space of flows already described decades ago by Castells (1996) and represent the present modes of social reproduction mechanisms.

In the next chapter we will attempt to disentangle the interdependence between digital and social structures, by exploring how revolutionary forms of DSI challenges technological determinism and the depoliticisation of urban governance deriving from delegating them to technological solutions (Morozov, 2012); and by questioning what some of the most pressing socio-political issues they raise in the urban space are.

References

Acuto, M., & Rayner, S. (2016). City networks: Breaking gridlocks or forging (new)lock-ins? *International Affairs, 92*(5), 1147–1166.

Akkerman, T., Hajer, M., & Grin, J. (2004). The interactive state: Democratisation from above? *Political Studies, 52*, 82–95.

Andrejevic, M. (2016). Theorizing drones and droning theory. In A. Završnik (Ed.), *Drones and unmanned aerial systems: Legal and social implications for security and surveillance* (pp. 21–43). Springer.

Ash, J., Kitchin, R., & Leszczynski, A. (2018). Digital turn, digital geographies? *Progress in Human Geography, 42*(1), 25–43.

Baccarne, B., Mechant, P., Schuurman, D., Colpaert, P., & De Marez, L. (2014). Urban socio-technical innovations with and by citizens. *Interdisciplinary Studies Journal, 3*(4), 143–156.

Barcelona Activa. (2018). *Mapping DSI—Cities and urban development*. https://digitalsocial.eu/images/upload/33-Cities%20and%20urban%20development.pdf.

Barcelona Digital City. (2020). *Barcelona digital city*. https://ajuntament.barcelona.cat/digital/en/blog/barcelonadigital-government-open-agile-and-participatory.

Bendiek, A., Godehardt, N., & Schulze, D. (2019). The age of digital geopolitics. *International Politics and Society*. https://www.ips-journal.eu/in-focus/chinas-new-power/article/show/the-age-of-digital-geopolitics-3593/.

Bourdieu, P. (1984). *Distinction: A social critique of the judgment of taste* (trans: Nice, R.). Harvard University Press.

Brabham, D. C. (2009). Crowdsourcing the public participation process for planning projects. *Planning Theory, 8*(3), 242–262.

Brenner, N., & Theodore, N. (2005). Neoliberalism and the urban condition. *City, 9*(1), 101–107.

Cammack, P. (2020). Marx on social reproduction. *Historical Materialism, 28*(2), 76–106.

Carayannis, E., Barth, T. D., & Campbell, D. F. J. (2012). The Quintuple Helix innovation model: Global warming as a challenge and driver for innovation. *Journal of Innovation and Entrepreneurship, 1*(1), 1–12.

Carlson, A., & Isaacs, A. M. (2018). Technological capital: An alternative to the digital divide. *Journal of Applied Communication Research, 46*(2), 243–265.

Castells, M. (1996). *The rise of the network society*. Blackwell.

Castells, M. (2008). The new public sphere: Global civil society, communication networks, and global governance. *The ANNALS of the American Academy of Political and Social Science, 616*(1), 78–93.

Chalmers, D. (2012). Social innovation: An exploration of the barriers faced by innovating organizations in the social economy. *Local Economy, 28*(1), 17–34.

Chesbrough, H. (2003). *Open innovation*. HBS Press.

Dacin, M. T., Dacin, P. A., & Tracey, P. (2011). Social entrepreneurship: A critique and future directions. *Organization Science, 22*, 1203–1213.

Davidson, M., & Iveson, K. (2015). Recovering the politics of the city: From the "post-politi- cal city" to a "method of equality" for critical urban geography. *Progress in Human Geography, 39*(5), 543–559.

Diaz-Parra, I., Roca, B., & Romano, S. (2015). Political activists' frames in times of post-politics: Evidence from Kirchnerism in Argentina and Podemos in Spain. *Contemporary Social Science, 10*(4), 386–400.

Dryzek, J. S. (2000). *Deliberative democracy and beyond*. Oxford University Press.

EC. (2020). *Next generation internet initiative*. https://ec.europa.eu/digital-single-market/en/next-generation-internet-initiative.

Galloway, A. (2004). Intimations of everyday life: Ubiquitous computing and the city. *Cultural Studies, 18*, 384–408.

Glimell, H. (2001). *The social production of technology: On the everyday life with things*. BAS Publisher.

Grassroots Innovation. (2020). *Researching sustainability from the bottom-up*. https://grassrootsinnovations.org/.

Greenfield, A. (2017). *Radical technologies: The design of everyday life*. Verso.

Grigoriev, S. N., Yeleneva, J. Y., Golovenchenko, A. A., & Andreev, V. N. (2014). Technological capital: A criterion of innovative development and an object of transfer in the modern economy. *Procedia CIRP, 20*, 56–61.

Hajer, M. (2003). Policy without polity? Policy analysis and the institutional void. *Policy Sciences, 36*, 175–195.

Hajer, M., & Wagenaar, H. (Eds.). (2003). *Deliberative policy analysis: Understanding governance in the network society*. Cambridge University Press.

Hamel, P. (2002) Conclusion: Enjeux institutionnels et défis politiques. In C. Gendron, & J. C. Vaillancourt (Eds.), *Développement durable et participation publique: De la contestation écologique aux défis de la gouvernance*. Presses de l'Université de Montréal.

Harvey, D. (2005). *Neoliberalism: A short history*. Oxford University Press.

Hodge, G. A. (2000). *Privatization: An international review of performance*. Westview Press.

Hofmann, J., Katzenbach, C., & Gollatz, K. (2016). Between coordination and regulation: Finding the governance in Internet governance. *New Media & Society, 19*(9), 1406–1423.

Ignatow, G., & Robinson, L. (2017). Pierre Bourdieu: Theorizing the digital. *Information, Communication & Society, 20*(7), 950–966.

Katzenbach, C., & Ulbricht, L. (2019). Algorithmic governance. *Internet Policy Review, 8*(4).

Latour, B. (2005). *Reassembling the social. An introduction to actor-network-theory*. Oxford University Press.

Lemke, T. (2002). Foucault, governmentality, and critique. *Rethinking Marxism, 14*(3), 49–64.

Lindberg, M. (2011). *Bottom-up development of innovation theory and policy*. Triple Helix IX International Conference, Stanford University, 11–14 July

2011, S1.1 History and conditions for success. Internet. http://www.leydesdorff.net/th9/Paper%20Malin%20Lindberg.pdf.

Lleialtec. (2020). Pla de Sobirania Tecnològica per l'acció comunitària. https://tec.lleialtat.cat.

Lynch, C. (2019). Contesting digital futures: Urban politics, alternative economies, and the movement for technological sovereignty in Barcelona. *Antipode, 52*(3), 660–680.

Lynch, C. (2020). *Networked sovereignties: Imagining alternative futures in Barcelona. Global Media Technologies and Culture Lab*. https://globalmedia.mit.edu/2020/07/01/networked-sovereignties-imagining-alternative-futures-in-barcelona/.

Mason, P. (2015). *Postcapitalism: A guide to our future*. Penguin.

Millard, J., & Carpenter, G. (2014). *Digital technology in social innovation, a deliverable of the project: The theoretical, empirical and policy foundations for building social innovation in Europe (TEPSIE), European Commission—7th Framework Programme*. European Commission, DG Research.

Misuraca, G., Pasi, G., & Urzi Brancati, C. (2017). *ICT-enabled social innovation: Evidence & prospective*. Publications Office of the European Union.

Mitchell, K. (2002). Transnationalism, neo-liberalism, and the rise of the shadow state. *Economy and Society, 30*(2), 165–189.

Morozov, E. (2012). *In the net delusion: How not to liberate the world*. Penguin.

Moulaert, F., Martinelli, F., Swyngedouw, E., & Gonzalez, S. (2005). Towards alternative model(s) of local innovation. *Urban Studies, 42*, 1969–1990.

Moulaert, F., & Van Dyck, B. (2013). Framing social innovation research: A sociology of knowledge perspective. In F. Moulaert, D. MacCallum, A. Mehmood, & A. Hamdouch (Eds.), *International handbook on social innovation*. Edward Elgar.

Mulgan, G. (2007) *Social innovation: What it is, why it matters and how it can be accelerated*. The Young Foundation.

Müller-Birn, C., Herbsleb, J., & Dobusch, L. (2013). Work-to-rule: The emergence of algorithmic governance in Wikipedia. In *Proceedings of the 6th International Conference on Communities and Technologies* (pp. 80–89). doi: https://doi.org/10.1145/2482991.2482999.

Parayil, G. (2005). *Information capitalism*. Springer SCOOP.IT. https://www.scoop.it/.

Pierre, J. (2000). *Debating governance: Authority, steering and democracy*. Oxford University Press.

Portney, K., & Berry, J. (2013). Civil society and sustainable cities. *Comparative Political Studies, 47*(3), 395–419.

Reichert, R. (2017). Digital governmentality. Citizen power, digital culture and city development. In G. Ziemer (Ed.), *Perspectives in metropolitan research* (pp. 36–46). Jovis.

Rhodes, R. A. W. (2007). Understanding governance. *Organization Studies, 28*(8), 1243–1264.

Rose, N., & Miller, P. (1992). Political power beyond the state: Problematics of government. *British Journal of Sociology, 43*(82), 173–205.

Sassen, S. (2004). Local actors in global politics. *Current Sociology, 52*(4), 649–670.

Seyfang, G., & Haxeltine, A. (2012). Growing grassroots innovations. *Environment and Planning C: Government and Policy, 30*(3), 381–400.

Shirky, C. (2008). *Here comes everybody: The power of organizing without organizations*. Penguin Press.

Smith, A., Fressoli, M., & Thomas, H. (2014). Grassroots innovation movements: Challenges and contributions. *Journal of Cleaner Production, 63*, 114–124.

Srnicek, N., & Williams, A. (2015). *Inventing the future: Postcapitalism and a world without work*. Verso.

Stoker, G. (1998). Public–private partnerships in urban governance. In J. Pierre (Ed.), *Parnerships in urban governance: European and American experience* (pp. 34–51). Macmillan.

Surowiecki, J. (2004). *The wisdom of crowd*. Doubleday.

Swyngedouw, E. (2005a). Governance innovation and the citizen: The Janus face of governance-beyond-the-state. *Urban Studies, 42*(11), 1991–2006.

Swyngedouw, E. (2005b). *Social innovation, governance, and community building—SINGOCOM*. Final Report Month 40, European Union Framework V, Key Action Improving Socio-economic Knowledge Base, Contract No. HPSE-CT2001-00070, Project No. SERD-.

Thompson, M. (2019). Playing with the rules of the game: Social innovation for urban transformation. *International Journal of Urban and Regional Research, 43*(6), 1168–1192.

Tracey, P., & Stott, N. (2017). Social innovation: A window on alternative ways of organizing and innovating. *Innovation, 19*(1), 51–60.

van Holm, E. J. (2017). Makerspaces and local economic development. *Economic Development Quarterly, 31*(2), 164–173.

Von Hippel, E. (2005). *Democratising innovation*. MIT Press.

Links

6AIKA. https://6aika.fi/en/frontpage/.

Ateneus de Fabricacio. https://ajuntament.barcelona.cat/ateneusdefabricacio/es/.

Barcelona en Comu. https://barcelonaencomu.cat/.

Decidim. https://decidim.org/.

Fab Academy. https://fabacademy.org/.

Infoshare. https://hri.fi/en_gb/.
Katuma. http://katuma.org/.
LOTI. https://www.londoncouncils.gov.uk/our-key-themes/london-office-technology-and-innovation-loti.
Small Technology Foundation https://small-tech.org/.
SocialHack Nou Barris. https://digitalsocial.eu/case-study/59/%7B%7Borganisation.url%7D%7D.
Som Energia. https://globalmedia.mit.edu/2020/07/01/networked-sovereignties-imagining-alternative-futures-in-barcelona/.
Transparent Krakow. https://przejrzystykrakow.pl.

CHAPTER 6

Power: The Raise of Critical Digital Social Innovation

Abstract This chapter explores how revolutionary initiatives of Digital Social Innovation (DSI) challenge the mechanisms of digital capitalism that generate systemic inequality *via* the implementation and functioning of digital urban technologies. By examining the "dark sides" and the potentially dystopic consequences of DSI, the chapter describes the struggle between competing ideological systems embedded in digital technologies as the new battleground for social justice. To this end, the chapter describes how cyber-activists, hackers and e-makers attempt at dismantling the complex entanglement of ownership, management and use of digital infrastructures and physical, economic, financial and social capital.

Keywords Digital social innovation · Cyber activism · Hacking · Digital capitalism · Digital participation · Big Tech · Small Tech · Digital revolution · Surveillance capitalism

6.1 Introduction

The birth of the internet and the *World Wide Web* dates back about 50 years ago when its creators intended it as a universally accessible means for people to exchange content and to collaborate (Sample, 2019).

C. Certomà, *Digital Social Innovation*,
https://doi.org/10.1007/978-3-030-80451-0_6

However, the possibility for commercial exploitation of the www soon became predominant with respect to its social vocation. Francesca Bria, President of the *Italian National Innovation Fund*, described this condition in the following terms: "A contradiction [...] exists at the heart of the Internet. Despite the existence of a technical networking layer that could spread power and give people more 'bottom-up' political and economic control over their lives, the existing commercial services built on top of this lower technical layer continues, for the most part, to empower existing 'top-down' centralised and established organisations in the corporate and government sector. It also often neglects smaller, and possibly game-changing innovative services aimed at tackling large-scale societal challenges" (Bria, 2015, p. 8). Despite being originally intended to subvert the XIX century elites and the disastrous effects of modern ideological politics (see Baricco, 2018), the digital revolution, in fact, has ended up creating a novel system of privilege and new power configurations, by overcoming the boundaries between traditionally opposite social categories (e.g. public/private, individual/society, rich/poor, owners/worker). The post-modern crumbling of dualistic categories of reference left room for the ideology of the free market, which also guided local policy agendas. One of the main challenges for humanity today is, thus, to understand and govern the digital revolution and its social consequences (Jenkins & Thorburn, 2003) against the neoliberal order backdrop.

When, in the last ten years, digital technologies massively infiltrated social life, it became necessary to ponder their socio-political and cultural impacts, especially in terms of (unfulfilled) promises of democratisation, empowerment and access for all. Notably, the transformation of the digital space into the terrain of conquest for multiple private interests that compete for the control of hard and software infrastructures, as well as the data flowing through them, is intrinsically ingrained within the mushrooming of DSI processes. Most of today's social narratives, collective objectives and political programmes materialise through different coding systems, software and digital platforms. Their creators advance diverse visions for the future, by vehiculating specific socio-political values connected with the political-economic value of the adopted technologies. This explains why the differences between diverse approaches to DSI (see Chapter 3) do not reside in "the what" projects are about, but rather in "the how and why" these projects are designed and promoted.

In this final chapter, we dig deep into the socio-political implications of technological solutions to understand how revolutionary DSI initiatives work as political gestures and (attempt to) address the issue of power in the digital age.

Criticalities (e.g. technical issues pertaining to the feasibility, effectiveness and efficiency of socio-technological processes) and critical issues associated with the diffusion of DSI initiatives (e.g. macropolitical issues of justice, empowerment/disempowerment, participation, inclusion, and ownership) have already been discussed by critical internet scholars (de Rosnay, 2006; Hunsinger et al., 2019; Stokes, 2020). Particularly, the role of digital corporations and financial institutions that act as ghostwriters of the local and global political agenda in the age of "digital feudalism" (Mazzucato, 2019) dominated by Big Tech companies, raises concerns about the actual freedom, transparency and independence of the public institutions from the market. Therefore, the shadowless narrative of digital technologies as emancipatory tools calls for a critical examination of the dark sides of the supposedly progressive, democratic and empowering character of DSI initiatives.

We start the chapter with analysing the intrinsic contradictions of DSI to explain why the technology-enthusiast rhetoric of the hyperconnected society have raised critical considerations on the potentially dystopic consequences (Söderström et al., 2014), such as forms of monopolistic appropriation, infrastructure control and power imbalances (de Waal, 2014), manipulation of citizens' opinions (Nielsen, 2006), (cyber) censorship (Loukis et al., 2017), and limitation of social dissensus pigeonholing (Caulier-Grice et al., 2012).

To defuse the hidden social technologies (see Chapter 3) that direct behaviour in a way which is functional to reproducing neoliberalist governance processes, radical and revolutionary forms of DSI intentionally challenge digital capitalism. Their effort is not only worthwhile for the macro-political consequences it might generate, but also for the micropolitics of the everyday digital, in which a complex entanglement between systemic inequality and digital space manifest in the different experiences with urban technologies (Keanbirch, 2019; McLean, 2020). Questions arise on how digital tools and processes embed political aims and social values (i.e. how technology choices encode socio-political values) and how the philosophy of cyber activism produces new imaginaries and practices of digital resistance. In the revolutionary DSI processes these

Fig. 6.1 A proclamation of digital discontent on the city walls. The text reads: "New chains are digital. Destroy them", Pisa, 2020 (photo: the author)

expose the augmented urban space manifests as the new socio-economic battleground (Fig. 6.1).

6.2 Contradictions at the Heart of Digital Social Innovation: Technological Fetishism and Digital Participation

In recent decades, critical geographers and urban scholars exploring the dark side of the augmented city gave clear insights to the fallacies DSI processes should avoid if they aim at working for the public good. Technological fetishism and digital participation represent two of the most debated issues.

The first one roots in the conceptualisation of the hyperconnected-city (see Chapter 3) as an information and communication machine that can be monitored in real-time and controlled at a distance (Batty et al., 2013; Kaika & Swyngedouw, 2000; Kitchin, 2014). This perspective produces the expectation that real and virtual components of the augmented city smoothly integrate, without friction or incoherencies, thanks to the implementation of urban interactive technologies. As a matter of fact, this integration is far from smooth. Vanolo (2013) explains that the expectation of a digitally managed city, created particularly through smart city programmes by ICT companies, "boosts the idea that technological networks and governmental practices will automatically guarantee better cities, regardless, for example, of the development trajectories of local societies, the nature of technological developments, the difficulty to reduce the chaos and complexity of ecosystems to a handful of statistics and indicators which have to be fully monitored and controlled, or the need for debates, rules and forms of control in order to achieve virtuous coupling between technology and society" (Vanolo, 2013, p. 896). When the hyperconnected-city project is understood as a mere matter of technological readiness, the commercial tension towards the scalability and replicability of innovations, discourages the creation of context-based processes in favour of one-size-fits-all solutions. This means that, in practice, the creation of the hyperconnected and (therefore) smart city, relying on the digitalisation of governance practices, encountered several problems and social oppositions (Fig. 6.2).

The technological fetishism associated with the hyperconnected-city narrative brings citizens to disregard that technical improvement of city infrastructures and data-driven solutions, although important, are not endpoints in themselves (Graham & Marvin, 2001; Sennett, 2012). The issue is not a minor one considering urban administrations often support DSI processes (see Chapter 3) that are functional to the interests of the technology-creator rather than of the public (see Chapter 2). Being able to discern who is truly benefiting from a new initiative and what macro-political visions this fits within, is therefore of major importance for mobilising digital technologies for the common sake—rather than for private ones. After the *Cambridge Analytica* scandal in 2018, which revealed fraudulent appropriation and manipulation of *Facebook* users' data, the perils for democracy represented by private ICT companies' harvesting people's data became apparent to the general public. These perils are particularly relevant in consideration of the tight entwining of digital-tech business and urban administrations.

Public opinion is progressively acknowledging that "the dynamics of power in the operationalisation of the smart city open the way [...] for normalizing the structural social violence" (Datta & Odendaal, 2019,

Fig. 6.2 The writing says "Ravage the Smart City" (photo: the author, 2020) expressing opposition to the smart city plan in Pisa, Italy

p. 387) of the tech industry. Critical movements are raising to protest, such as the *#BlockSidewalk* campaign in Toronto against the corporate capture of governance operated by the *BlockSide* company, sister of *Goole*; or the 2018 protest in Berlin against the opening of a *MountainView* campus under the slogan "Google go home!" (Fritze et al., 2019). Nevertheless, engaging with DSI in a critical way and rooting this engagement in the urban context is often very complex. Under the promise of broadening democracy and freedom through the technological upgrading of social reproduction processes, in fact, a depoliticised view of the city is vehiculated by the very same public–private partnerships that promote the hyperconnected-city model (see Chapter 4). Such a view hampers the critical reactions, resistances and the independent attempts of producing emancipatory forms of social innovation.

This brings us to the second issue, i.e. digital participation. Local governments largely deployed participatory processes with the aim of taking public voice into account in traditional decision-making processes, through a large variety of methods (Ledwith & Springett, 2010). Nevertheless, in the last few years, participatory practices have been charged with not being as inclusive and plural as expected and proved to have little impact on the enhancement of social justice and cohesion (Martinez & Rosende, 2011). They have been often intended, in fact, by policy makers as ways for consulting or engaging people in already existing projects and gaining their consensus in exchange for slight modifications to the original plans. The participation craze has been revived by the digital turn (Ash, 2009; Westera, 2012). The increasing availability of digital devices and software intended for knowledge-production and decision-making has been often welcomed, from both policy experts and social scholars, as bringing revolutionary possibilities for enlarging the sphere of public debate and democratising urban government and the governance processes traditionally precluded to most (Charalabidis & Koussouris, 2014). This condition stimulated the elaboration of urban imaginaries populated by intelligent cities smoothly planned, managed and controlled by highly integrated technological systems whose sentient and interactive character does not only allow (Kitchin, 2014), but also induces participation in other urban governance processes. DSI initiatives largely rely on the mobilisation of public participation. Citizens' involvement ranges from more passive processes, such as data provision and gathering, to more active forms of commitment exemplified by voluntary engagement of citizens in urban governance. However, the proliferation of participation venues and occasions in the digital urban space does not necessarily mean that real participation takes place (Davies & Simons, 2012; McLaren & Agyeman, 2015).

Critical scholars showed that practical, further than conceptual, issues might represent significant obstacles (Graham, 2014). Ozman and Gossart note that "there is an important gap between people's actual Internet use and public participation to open-source projects like Wikipedia [or to] citizen science, where participation to digital projects largely depends on social trends, such as the multiplication of well-educated individuals who do put their scientific knowledge to use in daily life" (Ozman & Gossart, 2018, p. 21). The translation of traditional participatory processes in the digital space raise doubts on the openness, accessibility and transparency of privately owned tools and the nature of public-private alliances promoting them, the new power geometries these foster and their empowerment/disempowerment effects (Baccarne et al., 2014; Bendiek et al., 2019; Parayil, 2005).

Furthermore, it is questionable whether user-generated urban governance processes (called "wiki-politics" by Cottica, 2010) are actually democratic. If we mean democratic as plebiscitary, it can be said that they are. However, if we endow the word democracy with qualitative content, we would have to admit that it is not necessarily so. In fact, participants select themselves on the basis of their skills (in particular on the basis of digital literacy) or shared values. Therefore, as Cottica observes, when wiki-politics constitutes the universe of politics and lets people autonomously create their own paths of access to public discussion, a scenario emerges that is totally different from plebiscitary politics (Cottica, 2010).

This consideration also urges us to reconsider the intrinsic value judgments associated with the idea of participation (Hague & Williamson, 2009), i.e. the more engagement, the better. The positive imaginary of participation to automatically produce greater democratisation, which is therefore inherently positive, is questionable (see Fig. 6.3). Participation is a value-laden but content-free buzzword which might, in some cases, legitimise, threaten or exclude initiatives that do not always increase the democratic quality of society. Despite that government-led projects might not incur such a problem, DSI initiatives (whose variety, multiplicator effects and often extemporaneous character can produce unexpected outcomes) run the risk of galvanising social groups whose agency (despite having been made possible by the existence of democratic conditions) do not necessarily create the conditions for democracy to persist. Moreover, the availability of another space for participation, i.e. the online public space, does not necessarily produce an increase in the number of participants, or in the quality of participation (see Fig. 6.3). Neither does it significantly help in differentiating the participants target nor in diversifying the participation modes (Baccarne et al., 2014, p. 7).

In general, despite digital participation's promise to democratise public debate, it does not address the potential negative effects for the quality of democratic debate, neither the causes of different levels of participation in diverse social groups. For instance, regarding quality decrease, the use of social media and the Internet for creating or subscribing petitions (called "clicktivism") while "increases the impact of a message by leveraging

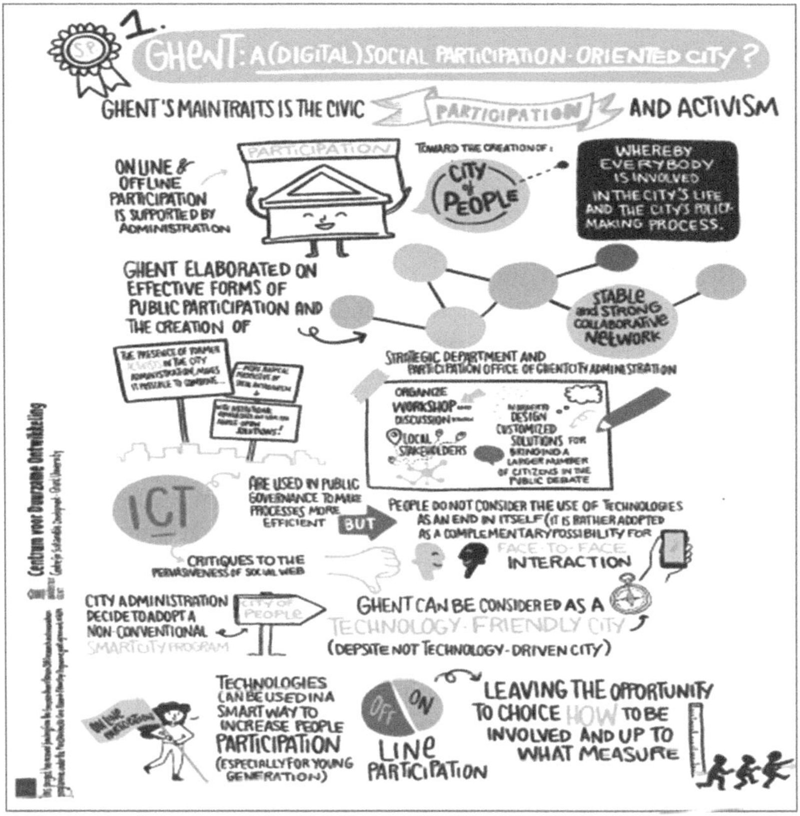

Fig. 6.3 An extract of a graphic representation of the of concerns about the meaning and practice of digital participation in Ghent, project EU MSCA project CROWD_USG (text by C. Certomà, graphic by V. Vitale, 2018; full visual story available at crowd.usg)

the Internet to further its reach [...] reduces activism to a mere mouse-click, yielding numbers with little or no real engagement or commitment to the cause" (Cambridge Dictionary, 2020). Evgenij Morozov warns against the disengagement effect of "clickactivism", which ends up with generating a detachment from the actual politics, rather than making citizens more involved in the public debate. With regard to differentiated levels of participation in digitally-mediated governance processes, the technological gap itself plays a key role in determining who participates and how they are involved (Cornwall, 2008). Consequently, a limited number of participants may end up representing the whole community (Guijt & Shah, 1998), and powerful interests may more easily determine asymmetries in decision-making processes (Mohan, 2001; Platteau, 2008).

Paradoxically, thus, despite claiming to enable open governance, DSI initiatives using participatory processes might end up having disempowering effects on a significant part of society and create new enclaves and power geometries characterised access and inclusion inequalities. Digital inequalities, however, do not only refer to differentiated access possibilities. Most of the people, especially in the global North have some possibilities of access to digital devices and to internet connectivity, but the quality of the contents (information, processes, services, personal connections…) they can reach is very low. As it is the case in food justice, already existing social, economic, political, educational, cultural disparities make most of the people only able to get junk digital contents. These problems immediately emerge when considering what technical skills are necessary to fully participate in the digital debate, and what policies and practical measures must be (and often are not) in place to allow full digital participation in the public life.

The issues of technology fetishism and contested digital participation signal how difficult is to engage for the public good *via* the promotion digital technology-based processes; and call for a more attentive consideration of the trivial but often ignored thevtruth that technology *per se* is never (politically) neutral.

6.3 Technology Is Never Neutral: Small Tech vs Big Tech

Technology neutrality in internet and data communication has been claimed as one of the key principles of *European Regulatory Framework for Electronic Communication* since 2002 (EC, 2009) and has been included among key pillars in the *OECD Internet Policy* (OECD, 2020). In these contexts, the term "neutrality" refers to design standards aimed at limiting negative externalities in the production of digital tools and processes. However, it is well acknowledged that "technology neutrality (or the absence thereof) can be used to nudge the market in a certain direction that is considered desirable by policymakers. For example, regulators might have a particular vision regarding the build-out of fiber networks. In order to implement that vision, the regulator may adopt rules that are not technology neutral" (Maxwell, 2014).

The claim that digital technologies are not neutral is generally advanced to contest the belief that technology is just a characteristic of certain tools that can be adopted to reach alternative and even opposite ends, depending on the users. In this perspective technologies (intended as technological instruments) are subject to the users' intentions. This is, however, a misleading perspective. Distinct technologies allow distinct functionalities and distinct modes of operation that are not (only) defined by their users. These have been originally planned by the designers and -even before- from the funders or the promoters of the initiative. And so, technology is not neutral from a socio-political perspective, not only because its use is not impartial, but also because they are created on the basis of specific assumptions and to accomplish certain goals (Bromley, 1998), which ultimately tend towards specific social configurations. Therefore, all phases of the living process of a technological tool, not merely its use, can be regarded as a value-bearer. This is particularly true for digital technologies in which the complex entwining of material and semiotic dimensions constitutes the base for their operational possibilities. As Zhao et al. clearly put it, "In reality, technology is neither passive nor neutral. It comes with shapes and expectations. A piece of software often conveys a certain teaching approach, which to a certain degree actively shapes what the teacher can do with it. [...] Ignoring the offers and constraints [...] of technology risks causing incompatibility between tasks and tools and can severely limit the effectiveness of the technology" (2002, pp. 24–25). The choice of tools can significantly impact the process outcomes because some digital tools are perfect for achieving certain goals but ineffective (or even counterproductive) to achieve others—and this is not just a problem for the project management, but it has ethical implications. For example, a digital platform aimed at gathering the care needs of the elderly people may be useful for caregivers but not for the elderly themselves who are not able to express their opinions by using the platform. Such digital divide problems are frequent in highly digitised urban services, which require particular attention with respect to who the urban public services are intended for and whether the tools are adequate for the purpose.

A further difficulty emerges in DSI processes when adopted software and hardware technologies are not ad hoc produced but are rather borrowed, extracted and re-arranged from existing infrastructures, tools or platforms. In these cases, the (crucial) technological components of a DSI initiative may not be designed for the specific initiative's purposes (for instance, for collecting citizens' opinions about the planning of a public park in a city). Consequently, when adopted, these technologies (implicitly and unconsciously) advance the vision and promote the interests of the company that created them. For instance, *Google Maps* allows citizens to annotate suggestions on a city administration's urban planning for free, upon the participants' agreement that these data are gathered by the company.

The digital product provider's vision and values are embedded within the adopted tool. When a tool is used, circulated and popularised, these values flow around with it. They shape the users' worldview accordingly—notably a worldview where market actors are entitled to allow, or not, citizens' access to the city's services based on the acceptance of the rules connected with the creation of their digital identities, that make them traceable and controllable. This generates reputational, as well as economic, benefits for ICT companies that provide the service or the tool, and qualifies them as legitimate interlocutors in institutional decision-making processes. In this way, the political vision of market institutions infiltrates and merges with public ones in promoting a specific urban vision. The hyperconnected-city imaginary, for instance, is fuelled by public-private partnerships and projects, such as the location of environmental wired sensors on the streets to monitor air quality or the realisation of a city dashboard for energy consumption monitoring. Therefore, private companies' intervention in urban government is legitimised and justified on the basis that the operational capacity of private companies is vital to public management (for example in providing adequate broadband coverage to make the use of the digital tool possible), and it also creates new economic value chains for the associated services. Such a value, however, is intended to fuel the business profit, rather than benefitting the public. And here is the main difference between Big Tech and the so-called small tech of e-makers and digital activists: whether the value generated by an innovation primarily benefits the company or society (still allowing the other to earn as collateral effect). The definition "small tech" is used by internet designers Aral Balkan and Laura Kelbag. It refers to digital solutions realised by independent small businesses, makers or associations whose intent is to make a change in society, by changing the digital tools it uses. There are various examples of this: the Balkan and Kelbag's *Small Tech Foundation*, a small not-for-profit company based in Ireland; another example is the *IndieWeb*, "a community of individual personal websites, connected by simple standards, based on the principles of owning your domain, using it as your primary identity, to publish on your own site [...], and own your data" (Small Tech Foundation, 2020); or the *Social Linked Data -Solid* project, led by Tim Barners-Lee, one of the inventors of the web, who aims to change the way the web works today, towards improved privacy, data-ownership and decentralised applications. These revolutionary digital social innovators do not work in isolation and often establish links with large international companies or public institutions, yet their goal is to work for the public and to transform processes of social (digital) reproduction. Balkan identifies the source of threats for democracy advanced by the Big Tech technocracy in the transformation of Internet from "the utopic vision of a decentralised and democratic commons [...] into the dystopic autocracy of Silicon Valley panopticons that we call surveillance capitalism" (Balkan, 2019b). He explains that "While it is true that surveillance capitalists crave to capture our attention and addict us to their products, they do this not as an end in and of itself but because the more we use their products, the more they can farm us for our data. Companies like *Google* and *Facebook* are factory farms for human beings. Their products are the farming machinery. They must provide a shiny façade to keep our attention and addict us so that we—the livestock—will willingly allow ourselves to be farmed. These institutions cannot be reformed. Big Tech can only be regulated in the same way we regulate Big Tobacco to reduce its harms on society. And we can—and should—invest in the ethical alternative: Small Tech". (Balkan, 2019b). Small tech, as synonymous for subversive forms of digital innovations for society, is presented as an antidote to the social diseases generated by the Big Tech because it is a copyleft technology, which is not built by for-profit corporations. Ideally it can be supported by the commons because getting a profit is not the primary aim of small tech tools, created to empower people and not to exploit them.[1] Digital innovators opposing the Big Tech model aim to bring back the internet to the

[1] In fact, people's data stays on their devices or are end-to-end encrypted "and the individual who owns the tool has sole possession of the keys to their private information and sole control over who the ends" (Balkan, 2019b). This commitment to privacy is coupled with a simple and modular design approach so that the realised tools can be

people, as well as their data (and ideas), which are generally appropriated from corporations.

6.4 The Data Issue

The dismantling of Big Tech techno-capitalism primarily requires the creation of an open and decentralised Internet to serve the people, rather than the profit of the developers; and an ethical management of the main contemporary tradable assets, i.e. people's data, which should be managed jointly through cooperative ownership systems.

Big Tech companies build their economic success by leveraging and exchanging people's data that is provided for free, because it isn't even recognised as a value or currency. Through a combination of big data analytics, ICT companies know more about users than we imagine. Kelbag explains that "Recommendation engines are used by companies like *Netflix* to deliver great personalised experiences. *Google* creates our profiles to understand what drives us to sell us the right products. *23andme* analyzes our DNA for genetic risk factors and sells the data to pharmaceutical companies. Ecommerce sites like *Amazon* know how to attract you as an individual [...] *Facebook* can predict the likelihood of you drinking alcohol or drugs, or determine if you are physically and mentally healthy. It also experiments on us and influences our emotions. This data is of tremendous value to people who may not have your best interests at heart" (Kalbag, 2015). The issue is generally discussed in terms of privacy, but such a perspective is somewhat misleading because many people seek a greater visibility rather than greater protection of privacy in a world where popularity is measured by views and likes. On the contrary, Balkan comments: "we must resist any attempt to reduce people to property with maximum of fervor. Because not doing so means giving our tacit consent to a new slavery: one in which we do not trade in the biological aspects of human beings but in their digital aspects" (Balkan, 2017). Tim Barner-Lee describes in very clear terms the risk associated with the loss of personal data: "Since our data is [...] stored in proprietary silos, out of our sight, we lose the benefits we could achieve if we had direct control over this data and chose when and with whom to share it. Furthermore, we often have no way to provide companies with which data we would

"understood and maintained and improved upon by the greatest number of people" possible (Balkan, 2019b).

rather not share, especially with third parties, the T & Cs [i.e. *terms and conditions*] are all or nothing" (Barner-Lee, 2017). This situation also has a direct impact on governance, because through public-private partnerships, both companies and governments gain increasing control over data, which can have harmful consequences, especially in repressive regimes. It is difficult to even imagine what companies can do in terms of social experimentation and social orientation thanks to the enormous amounts of data on billions of users collected every time a proprietary service is used (obviously under explicit consent given when users agree on terms and conditions—although very few people are able to understand them or bother to read them). For example, all social media users are already unknowingly taking part in thousands of experiments and every time we access our accounts someone makes a profit. The problem is not just having tools or service providers that respect the privacy of the users, but also finding more effective alternatives when designers of the companies put pressure on users to agree on all the conditions of use in exchange for the service provided. This is why Berners-Lee's *World Wide Web Foundation* suggests partnering with companies to create alternative revenue models, to fight the government's overrun in surveillance law, and fend off disinformation by engaging Big Techs to increase algorithmic transparency (DW, 2019).

Furthermore, the data issue can also be addressed in terms of social justice. Some private companies (most of them located in China and the U.S.) are milking data from individuals and social organisations to provide internet-based services (apparently for free) in exchange for their consent to disclose information about their research, activities and profiles. The big idea is to profit from something that most people in the world give little or no value to. Data are the new oil to extract from the crowd. These increase gain for the data-owning companies without any benefit for society as a whole because the nature of most ICT companies allows them to pay a limited amount of national taxes and make large use of underpaid and precarious work. However, these are just some of the fallacies, pitfalls and criticalities that digital activists are confronted with and to which they attempt at providing alternative solutions for, while at the same time tackling pressing social problems through creative innovations. Bigger companies also hold the technological know-how and have the financial capacity to acquire all emerging start-ups, which produce innovative technical components, including those engaged in DSI initiatives. By doing so, they assume a quasi-monopolistic position that is reinforced by agreements with companies that own or manage communication infrastructure, and that is deterrent to the emergence of alternatives.

What does all this have to do with the DSI initiatives taking place in European cities? When public administrations undertake the processes of digitisation of government and governance processes, they generally entrust the technical aspects to ICT companies. In addition to being paid for the service, they also have the opportunity to acquire citizens' data for free. At the same time, the political, social and symbolic power these companies gain from collaborating with the public allows them to influence decision-making processes and direct city development plans. It is not uncommon, therefore, that the DSI initiatives supported and strengthened by local administrations operating in agreement with large ICT companies are only those that create a market value (rather than social) and that are functional to the reproduction of existing conditions. Alternative DSI initiatives find it very difficult to emerge, resist and promote forms of public governance that run counter to the mainstream neoliberal model. The overall issue is clearly highlighted by Traldi and Bria in the following terms "Who controls data, AI and digital infrastructures will determine the nature of future institutions. To maintain European social models and defend values and rights, citizens must hold reins of technology" (Traldi & Bria, 2018). One such attempt was made, for example, by the *Strategic Department* of the Ghent city council headed by EU consultant Karl-Filip Coenegrachts a few years ago. He promoted the creation of a free, open, decentralised and interoperable personal federated website for the city "to create a new kind of Internet [...] where people control their data, it is not tracked and everyone has a space equal online" (Harris, 2018). It is therefore possible to break the Big Tech monopoly by requiring companies to be transparent, pay taxes, respect ethical principles both in their production processes and in the operation of their products. This is possible when, as in the case of Ghent, the initiatives of the DSI are taken as an opportunity to create a viable environment for alternative processes to the existing ones. As Coenegrachts claimed during the European Parliament conference *Saving the Internet for the Many, Not the Few* in 2019: "Only by decentralizing [...] data, technologies and the Internet in general, our cities will remain places where people themselves they are in control" (Coenegrachts, 2019).

But the ridge is steep. The new *Future Internet Public–Private Partnership* programme approved by the European Commission itself ambiguously promotes a user-centred but industry-driven vision of European Internet technology governance (EC, 2013). Journalist and activist François Soulard summarises the problem as a question of combining geometries of power: "the Internet governance model resembles above all a transversal and multisectoral geometry. [...] It is a model that must fundamentally deal with the relationship between scales, actors and thematic issues, combining different modes of action (multilateral dialogue, sovereign decision-making, co-production of norms, multi-sectoral participation, subsidiarity of civil and commercial law, etc). This complex geometry is new and disturbing, both in terms of political practice and theory" (Soulard, 2018). In this context, the strategic and political role of cities in tracing the geometries of Internet governance and regulating them is made clear by the *Cities for Digital Rights Declaration*. Jointly promoted by some 40 cities around the world in support of the United Nations Human Settlements Programme, the *Declaration* is based on the belief that cities are the democratic institutions closest to people. At the same time, they are also the most advanced front in the battle "to remove obstacles to the exploitation of technological opportunities that improve the lives of our constituents and to provide reliable and secure digital services and infrastructures that support our communities" (Cities for Digital Rights Declaration, 2020). Therefore, the participating cities undertake to respect the principles of equal access, transparency and responsibility, non-discrimination of data and algorithms, participation and inclusion, protection of privacy, freedom of expression and democracy in the "design into digital platforms starting with locally-controlled digital infrastructures and services" (Cities for Digital Rights Declaration, 2020). Despite some vagueness in identifying responsibilities for the misuse and socially negative consequences of digital technologies, the *Declaration* clearly highlights the role of cities in this fight globally, which has often been conceived as something de territorialised. This point recalls the affirmation of geographers Joe Shaw and Mark Graham: "Cities have become more than bricks and mortar: they are their digital presences, and they are constantly represented and reproduced as such. [...] If our cities are now digital, then the struggle for more egalitarian rights over the city must go beyond just focusing on material spaces and in the digital realm" (Shaw & Graham, 2017, p. 4).

6.5 Building Up Alternatives: Philosophy (and Practice) of Revolutionary Digital Social Innovation

In line with Shaw and Graham's intuition, digital social innovators inspired by the subversive intent to unveil the socio-political positioning of technologies, take stance against the super-power of the Big Tech. They operate in the augmented urban space, because the digital dimension is the new battleground where the struggle for a more equal, democratic and inclusive society is fought (Hunsinger et al., 2019). Their agency is inspired by the culture of hackers, digital makers and cyber activists, characterised by a deep commitment for information freedom, a pervasive mistrust for authority, dedication to meritocracy and the belief that computers can make beauty and better the world (Coleman, 2013). The digital anthropologist Gabriella Coleman extensively explored the hacking culture and defined it as the most important phenomenon in "global culture and politics in the late 20th and early 21st century" (Coleman, 2020). She explains that the existence of hacking culture has been acknowledged from 2007, when the Anonymous movement played a key role at *WikiLeaks*, the Arab Springs or *Occupy Wall Street* protests (Coleman, 2014), unveiling the social hypocrisy and dogma behind the fake neutrality of technology. The movement per se is, however, difficult to identify because its nature is anarchist and ambiguous: "Hackers do different things and cohere into distinct traditions. They are people of different kinds: they are young and old, men, women, and nonbinary. They are not invisible, immaterial, hidden or cloaked except, of course, when they want to be. Hackers are funny, and scary; hackers are do-gooders, and assholes; they write code; hunt for bugs; tell the government to fuck off; land in jail (and even break out of jail); they work for three-letter agencies and expose them; they secure and break into our systems; invent new laws" (Coleman, 2020). Therefore, the hacking culture is not a coherent, manifest-based and clearly identified one; rather is a way of being that presupposes the capacity to infiltrate in the ganglia of the socio-technological system and operate on its very functioning. Nevertheless, the hyperbolic traits of the Anonymous movement made it easier for everybody to identify the culture of internet activists as an attempt at bringing back digital technologies and the internet to the people (see Turner, 2006). It roots in the vision of the fathers of the digital revolution (such as Steward Brand with his *Whole Earth Catalogue*) that have sown the seeds of the cyber-counterculture and "offered a vision for a new social order—one that eschewed institutions in favour of individual empowerment, achieved through the acquisition of skills and tools [; for] its irreverence toward institutions, its emphasis on autodidacticism, and its sunny view of computers as tools for personal liberation" (Wiener, 2018).

Although the digital revolution was intended to subvert the XIX centuries' elites and to redistribute opportunities to people (Cadwalladr, 2013; Turner, 2006), after a few decades, it produced a massive concentration of economic, financial and political power in the hands of a few big IT companies. These companies can acquire patents, engaging hackers and invest in all of the promising IoT and AI initiatives. From the disruptive intents of the cyber-counterculture, the dark sides of internet culture emerged and the digital turn ended up with producing its own elite. As Brand predicted: "A society of large tools cannot be democratic, egalitarian, socialistic, humane, and just. It must be hierarchical, exploitative, bureaucratic, and authoritarian. If the day comes when all of humanity's wants can be supplied by a few giant tools, the people who tend them will rule the world" (Streshinsky, 2018). Most of the financial and economic capacity of companies that are supplying these "giant tools" derives from three favourable conditions: they trade in personal data which users voluntarily provide in exchange of access to services and devices apparently provided for free; they often pay a very limited amount of nation-based taxes because the new supranational business model of giant ICT companies make them able to overcome the nation-based rules; and they gain from the immense traffic of news, information, and ideas they advertise and control. Therefore, to restitute the internet to the people it is necessary to erode the fake narrative of speed, freedom/liberty and accessibility upon which the Big Tech corporations build and feed their empire (Kalbag, 2017). Inventing new ways for making the internet a common good has been in the air for quite some time, but it encounters several difficulties. Soulard summarises them as follows: "The internet is basically a network of information networks that allows the exchange of information between computers *via* a shared protocol: the TCP/IP protocol. It is also a complex system, in as much as it constitutes in itself a web of interconnection of socio-technical subsystems, where local, regional and global models of infrastructures, uses and content are juxtaposed. Subsequently, resources mobilised in the digital sphere are by definition combined and plural. There are mixed resources (assignation of domains, exchange points), public resources (energy, digital services), shared resources (protocols, standards, norms, servers, open code, content), and private resources (transoceanic fibres, data centres, proprietary code, content). Regarding the nature of these resources, commons specialists do not narrow the digital environment down to strictly one common good. For them, it is rather a common pool resource, a hybrid compound of shared resources. In this context, the notion of common good, applied to the internet, refers to a regulatory perspective or aim" (Soulard, 2018). Such a complex entanglement of public-private components characterise the internet and make it difficult to treat it as a common good. Mark Raymond, consulting at the think-tank *Centre for International Governance Innovation*, explain why this is the case. Being, in fact, an extremely open domain with sparse rules for its members, the internet has been mistaken for a common good, however, from a technical perspective, it is a rivalrous good (because the computer networks on which it depends accommodate a finite amount of traffic). In fact, even though more fibre optic cable, switches and routers can be constructed, or more efficient protocols for routing and directing traffic can be designed, the possibility for the internet to be regarded as non-excludable good is weak for three reasons. First of all, Internet service providers in many places of the world are required to block some content (such as hate speech or child pornography). Secondly, distributed-denial-of-service attacks accomplish short-term exclusion by bombarding a targeted website with requests for information, and this can have the unintended consequences of denying access to additional, unintended targets. Third, because of the terrorism threat, it is possible to exclude people from the Internet by destroying a part of the physical infrastructure (i.e. fibre or wireless) (Raymond, 2012). Nevertheless, Raymond recognises that, whether a common or not, it is essential to maintain the global reach and interoperability of the Internet to ensure that access is open to all (aside minimum restrictions for public safety requirements) in order to maximise the value of the internet for humanity (Raymond, 2012). It is exactly with the aim of making the internet as public (and publicly accountable) as possible that the vast commons movement in Europe has connected with the peer-to-peer movement. This is a type of social relationship that is non-hierarchical and non-coercive, taking place in human networks that can be scaled up and widespread by technological infrastructure. It allows computers in a network to interact without going through a separate server, but with specific orientation, so that further than allowing participants to connect and organise in a network, they can independently define the purpose of their collaboration. The relationships between peer-to-peer and the commons movement (described, amongst others, by the commons theorists Michael Bauwens, author of the *P2P Manifesto*) creates the conditions to optimise the use of existing resources in common and the rule for this use (Commons Network, 2020). However, the realisation of the cyber activists' utopia of a city-based interoperable infrastructure, through which data is held in common and managed by the people, requires a significantly different digital architecture from the existing client/server one. Such an ideal internet space can only be funded by the commons and would serve as a permanent node within a peer-to-peer structure. This attempt is resonant with the original conception of the www as "an open platform that would allow everyone, everywhere to share information, access opportunities, and collaborate across geographic and cultural boundaries. In many ways, the web has lived up to this vision, though it has been a recurring battle to keep it open" (Berners-Lee, 2017).

Such a battle is today an integral part of the more subversive and revolutionary manifestations of DSI aimed to change the current functioning of the web. This is conducted by several innovators that share "a deep dislike of large concentrations of power and a belief in the kind of egalitarian, pluralistic ideas they say the internet initially embodied"; and that operate against the digital industry "grown greedy, intrusive and arrogant – as well as governments whose surveillance programmes have fuelled the same anxieties" (Harris, 2018). Some of them, such as people working on the *MaidSafe* team want to rethink the www architecture to crumble the Big Tech monopoly. For instance, their *Safe Access for Everyone SAFE* project envisages the creation of a network in which "further than being stored on distant servers, people's data—files, documents, social media interactions—will be broken into fragments, encrypted and scattered around other people's computers and smartphones, meaning that hacking and data theft will become impossible. Thanks to a system of self-authentication in which a SAFE user's encrypted information would only be put back together and unlocked on their own devices, there will be no centrally held passwords" (Harris, 2018). Obviously, there are major problems in introducing alternative modes of using and regulating the internet, which seems not to have an easy solution. For instance, the payment of services, or the fact that information stored on people's devices will potentially include nasty or illegal materials which, if encrypted, cannot be controlled to prevent damages to other people.

Again, from the users' perspective, the social experience one can get on existing decentralised internet and social networks (e.g. *Steemit, Diaspora* or *Mastadon*) offer no incentives to go because they are largely under populated at the moment (given the strong concurrence of the pop social media, such as *Instagram, Pinterest, Twitter* or *Facebook*) (Banning, 2015). However, Balkan suggests that to appreciate the success of these alternative initiatives we need to reject the evaluative approach adopted in venture capitalism initiatives, whose success is assessed on their capacity to vertical upscaling. Rather we need to focus on their capacity to forge horizontal cooperative networks that interoperate with each other without the intermediation of major for-profit companies. The point is not abolishing big companies (which are very law protected) but obliging them to behave ethically.

6.6 Against Surveillance Capitalism

We already considered (see Chapter 2) how the city represents the context for experimenting local, revolutionary and grassroots-based DSI processes that have the potential to overcome the mainstream approach to innovation which is functional to the reproduction of neoliberal digital capitalism(Fig. 6.4)

In fact, when public institutions collaborate in revolutionary digital social innovators' projects, a number of alternative solutions to those created by the Big Tech companies can be produced and upscaled. Rather than adopting market-oriented products, people-oriented ones can be funded without the intervention of venture capitalism and (the part of) the private sector that does not care for peoples' rights.[2] Digital designers and developers engaged in producing ethical and people-oriented technological products are often very poorly economically and politically supported. Their solutions are scarcely acknowledged and adopted by institutions. The only option for many is to sell their products to the Big Techs and therefore to feed the monopolistic system of "surveillance capitalism". Surveillance capitalism, corresponding to the modus

[2] For instance, technological alternatives include apps developed by small, independent and not-for-profit companies, such as *Fastmail* for emailing, Signal or Wire for messaging, *Mastodon*; web construction set such as *Site.js*; or also hardware such as *Pinephone*, which uses free and open hardware products. *Switching.Software* provides a list of ethically built, privacy-conscious alternatives to well-known software.

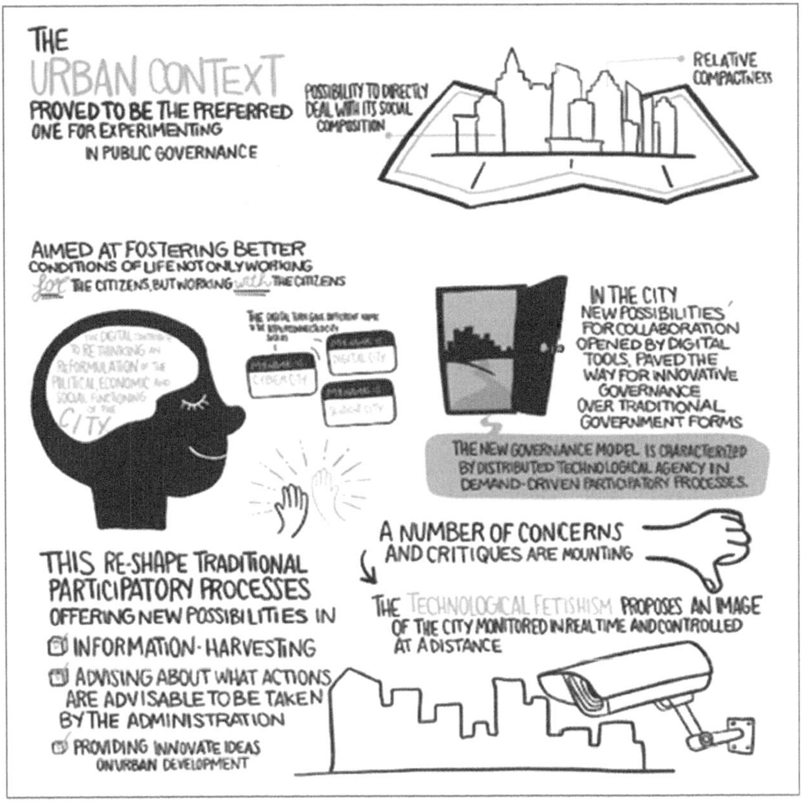

Fig. 6.4 A graphic representation of the considerations the city's appropriateness for experimenting in DSI, project EU MSCA project CROWD_USG (text by C. Certomà, graphic by V. Vitale, 2018; full visual story available at crowd.usg)

operandi of the Big Techs, works by giving access to billions of people free online services but enables the providers of those services to monitor their behaviour. Exposed by Shoshana Zuboff, it is described as follow: "Surveillance capitalism [...] unilaterally claims human experience as free raw material for translation into behavioural data. Although some of these data are applied to service improvement, the rest are declared as a proprietary behavioural surplus, fed into advanced manufacturing processes known as 'machine intelligence', and fabricated into prediction products that anticipate what you will do now, soon, and later. Finally, these prediction products are traded in a new kind of marketplace that I call behavioural futures markets. Surveillance capitalists have grown immensely wealthy from these trading operations, for many companies are willing to lay bets on our future behaviour" (Naughton, 2019). Digital capitalism, therefore, divides society into observers (invisible, unknown and inexplicable) and observed ones. The whole system is designed "to evade individual awareness, undermining human action, eliminating decision-making rights, decreasing autonomy and depriving us of the right to fight. The big picture reveals extreme concentrations of knowledge and power" (Zuboff, 2019). This condition reinforces existing asymmetries (and produces new ones) and can only be countered by a global initiative to regulate the functioning of ICT companies, because, as Zuboff recalls: "Surveillance capitalism thrives in the absence of law. We have not failed to curb this rogue capitalism; we have not yet tried" (Zuboff, 2019). The cogency of such an intervention is justified by the consideration that even if people acquire ownership of the data, they supply a company, they do not own the resulting predictions and even less do they own the market revenues from their sale. A collective and systemic legal intervention is crucial to disrupt and outlaw the data provision to surveillance capitalism and the resulting income streams. On the other hand, Zuboff suggests that laws and regulations designed to benefit companies that want to break with surveillance capitalism must be created and that lawmakers must join new forms of collective action to achieve these ends (Zuboff, 2019). Overall, the problems of surveillance capitalism are collective, social, non-technological. It needs political decisions to allow the emancipating forces of digital innovation to work for the benefit of the public.

Confronted with the deafening silence of international organisations and States on the massive economic power of high-tech multinationals, in many cases, cities stand out as the bulwark of people's protection against the logic of profit. Their possibility to counteract the digitally enabled neoliberalisation of social services is very limited and keeping the results safe is a difficult task, as the following example shows. The EU-funded *DECODE* project is aimed at providing tools that put individuals in control of their personal data, deciding what to share and for what purpose. Lead by the cities of Amsterdam and Barcelona, the project proposed to reinforce citizens' digital rights by "building a data-centric digital economy where data that is generated and gathered by citizens, the Internet of Things, and sensor networks is available for broader communal use, with appropriate privacy protections" (DECODE, 2020). In this way innovators, start-ups and citizens' associations can take advantage of that data and use them in a controlled and democratic way. However, Balkan named the DECODE project as a further and eminent example of how EU-funded research and development feeds Silicon Valley's revenues. He claims that "We [i.e. European citizens] fund startups and, if they're successful, they get bought by the Googles and Facebooks. If they're unsuccessful, the EU taxpayer foots the bill. It's time for the European Commission and the EU to stop being useful idiots for Silicon Valley, and for us to fund and support our own ethical technological infrastructure. [...] This isn't to say that we must wall ourselves off or create a European silo. On the contrary, we must ensure that the technological infrastructure we fund and build is free and open, decentralised, and interoperable so that anyone, anywhere in the world can use it" (Balkan, 2019a). If such conditions were already in place, it would have been impossible for public-funded research projects to serve private enterprise, like for instance *DECODE* supporting the research of *Chainspace* (a London-based blockchain venture) to then be acquired by *Facebook* for the creation of its cryptocurrency programme *Libra*.[3]

The way forward is therefore disseminated of pitfalls and obstacles, but there are no other ways for creating tomorrow's digital space for the people, rather than for corporatocracy's profit.

6.7 Conclusion: Digital Resistances

Previous chapters described how in the early 2000s the digital turn determined a proliferation of web-based processes that granted existing social innovation initiatives with an unprecedented possibility to increase their efficiency, diffusion and effectiveness. By widening public access to decision and policymaking processes, and harnessing the digital connectivity potential to offer disintermediate services, DSI is today playing the trump cards of the digital revolution. This is having a disruptive impact on contemporary cities, which serve as laboratories for one of its most exemplary manifestation, the DSI. From enabling citizens' participation in

[3] Even the magazine *Wired* noted that: "Ironically enough, the work of an EU-backed company working on decentralisation and privacy caught the eye of EU-fined data-guzzling monopoly Facebook" (Volpicelli, 2019).

decision-making processes, to providing new social services, or promoting people's creative potential in urban planning, DSI initiatives are proliferating and transforming contemporary worldwide cities. The city, in fact, seems to offer the necessary conditions for experimenting with DSI, by virtue of its relative spatial compactness, social density, connectivity and proneness to test decentralised solutions to social challenges. At the same time, the diffusion of digital technologies, the interactive web and the IoT are modifying the physical structure and the functional organisation of the city (Komninos et al., 2012). With the digital turn both the concept and the experience of urban space underwent profound changes, notably due to the merge of the physical and the virtual dimension into a single "augmented urban space". Novel collective imaginaries and narratives emerged from the encounter between society and space mediated by digital tools and processes (De Souza, 2006); and generated distinct digital spatialities (Sheppard, 2004). Digital technologies continuously re-construct the city by generating social representations, influencing perceptions and directing the flows of money, resources, people, and circulating ideas. Therefore, while dismissing many of their political and economic responsibilities as mere technical issues, digital technology providers are actually acting as one of the major drivers in the fabrication of augmented cities. Notably DSI is not only to be considered as the application of digital tools to existing social innovation processes but rather, it makes evident that a new social technology has emerged and is shaping peoples' lives.

The original trait of DSI—compared to traditional SI—in fact, resides in its ability to modify the organisation, the functional processes and even the understanding of society and space relationships by operating changes in the digital domain. The DSI domain, however, is not flat and radically different approaches coexist. These range from the market-led and technocratic approaches to innovation, embedded within the digital capitalismsystem; to the reformist one promoted by the European communities of digital social innovators imbued with the social-democratic values; up to the radical digital activists who work outside institutional settings to re-frame the current socio-technological cognitive paradigm, subverting the system of digital governance, and dismantling surveillance capitalism. In general, thus, DSI is a contested concept that conveys the most critical elements of the digital revolution and represents one of the major forces driving contemporary social development, that—in reason of its fluid, ephemeral and fragmented nature—has been too long neglected by social scholars. Most existing research on DSI, in fact, has been supported by the EC *via* the *Research and Innovation* funding schemes. Moreover, a diversified, vibrant and extremely rich panorama of DSI initiatives independently promoted and self-financed by city councils, NGOs, CSOs or social entrepreneurs has been documented up to now. The relationships of DSI with the urban context have been, however, only marginally explored. This lack is likely due to the dominance of the innovation management approach, which fails to account for the pitfalls and the "dark sides" of the digital revolution, and the space-related social, cultural or political effects of DSI practices (including the creation of polarised power geometries, the novel social injustices and technology imbalances).

By adopting a critical geography perspective, we explored the spatial conditions where DSI projects take place and the social geography they entail, with particular reference to the issues of social reproduction (i.e. how DSI practices reproduce themselves and the city), representations (i.e. how the symbolic production of DSI entwines with the urban narratives) and power (i.e. how DSI fuels and flows through power geometries and geographies). We considered how the very spatiality of the city becomes transformed by and through DSI.

Digital technologies are today ubiquitous, and the internet has created a truly global public space, dreamt of by cosmopolitanist philosophers, where local and individual issues can be exposed and debated. Nevertheless, after the first enthusiasm, it became clear that several theoretical and practical issues came along with the apparent free, democratic and participatory character of the internet. Mark Graham explains that "the Internet is not an amorphous, spaceless, and placeless cloud. It is characterised by distinct geographies. Internet users, servers, websites, scripts, and even bits of information all exist somewhere. These geographies of information shape both what we know and the ways we are able to enact, produce, and reproduce social, economic, and political processes and practices. [...] Even in an age of almost ubiquitous potential connectivity, online voice, representation, and participation remain highly uneven" (Graham, 2014, p. 17). These uneven digital geographies clearly emerge when we consider where information, tools and contents are produced and made available on the internet, and this determines new geometries of political and economic dependence (Graham, 2014). For instance, many people, even in Europe, are still out of the global interconnected network because several elements contribute to the creation of the digital divide, including "interrelated issues of literacy and education; digital architecture; physical infrastructure; governance of online communities and platforms; cultural, religious, gendered, and other socially constructed barriers; politics and political interference; and language" (Graham, 2014, p. 31). All of these issues require a systemic approach to understanding and governing the socio-political implications of the digital, stepping beyond the acritical optimism for technologies and paying attention to the critiques elaborated by the discontent of the "wisdom of the crowd" mantra (Lanier, 2006).

Whilst advertised as a panacea for handling global governance complexities, DSI can serve neoliberal plans and anarchic initiatives. The heterogeneity of the urban imaginaries (from the hyperconnected, to the receptive and do-it-yourself city) DSI practices are associated with justifying the mobilisation of diverse—often opposite- approaches, ways of performance and practices. By exploring them, we can contribute to a political analysis of the digital turn; to question the socio-political underpinning of traditional digital governance; and to disclose the hidden, uneven geographies of the digital space that create spaces of empowerment and disempowerment, privileges, and exclusions.

We cannot, thus, treat DSI as a single phenomenon or stigmatise it as an offspring of the digital corporatocracy. This would be reductive and not give an account of the myriads of efforts by revolutionary activists that are attempting to change from inside the structures and functioning of digital capitalism. In urban contexts, digital social innovator communities are fighting, code after code, the struggle between competing models of digital evolution. In doing so, they are moving the social struggle in the digital space battleground; and are repositioning this last in the city.

Their practices signal the distortive effects of digital technologies on the political life of a society, which are multiple and range from click-activism to the more consequential and systematic avoidance of critical discussion on the very heart of the digital revolution, including issues of ownership, privacy, access equality, security, real freedom, power imbalances and system failures. The struggle between competing ideological systems embedded in digital technologies and the underlying socio-political visions advances in multiple directions. This connects with the concerns for who should be responsible for imbuing public values in digital innovation and guarantee that conditions of fairness, accessibility, democratic control, and accountability are respected for both users and workers of the digital economy. At the same time, this struggle unveils the socio-spatial effects of the adoption of digital technologies in urban governance, in which predictive analytics are used as a social technology for the measuring of socio-economic dynamics and for establishing novel digital governmentality.

Against the backdrop of emerging and mutable geometries of power, which emerge and disrupt existing urban governance routines, critical innovators are operating in a technical domain, which is not accessible or understandable to the most. They attempt at changing the rules of the game by entering into and dismantling the complex entanglement of ownership, management and use of digital (hard and soft) infrastructures, and physical, economic, financial and social capital.

What the existing attempts at advancing revolutionary DSI processes in the urban context tells us is that building digital technologies in a way that eschews the consequences of surveillance capitalism goes together with a vision of the "city of people" (Connegracht, 2020) that mobilise technologies for the common good. As the *Shared Digital Europe Manifesto* by *Commons Network*, drafted and signed by many European digital activists, claims, the digital space is at once the effect, the frame and the essence of our society, therefore it cannot be simply defined as a place where only economic dynamics count. This would generate the misleading perception that the digital space is only a marketplace, in which public institutions and civil society have a marginal role in supporting one or another private venture. Rather, the *Shared Digital Europe Manifesto* embraces democratic values and strives for equity and social justice *via* enabling processes of self-determination on private data, the cultivation of the commons, decentralisation of infrastructure and empowerment of public institutions (Communia, 2019). Specific attention is required for how data are managed and mobilised in building up a city in which maximisation of the efficiency of services for society goes together with the attention towards (digital and non-digital) commons, and with a transcalar view, that frames the local decisions in a global perspective.

This would helps us dismantle the great expectations of digital participation as a panacea and suggests the way for a more context-based appreciation of the barriers preventing the realisation of the promises of the digital revolution (including democratisation, wider accessibility to data, information, opportunities, connections, and similar) and subsequently the empowerment, efficiency and effectiveness of governing systems.

References

Ash, J. (2009). Emerging spatialities of the screen: Video games and the reconfiguration of spatial awareness. *Environment and Planning A, 41*(9), 2105–2124.

Baccarne, B., Mechant, P., Schuurman, D., Colpaert, P., & De Marez, L. (2014). Urban socio-technical innovations with and by citizens. *Interdisciplinary Studies Journal, 3*(4), 143–156.

Balkan A. (2017). *Encouraging individual sovereignty and a healthy commons*. https://ar.al/notes/encouraging-individual-sovereignty-and-a-healthy-commons/.

Balkan, A. (2019a). *Have you heard about Silicon Valley's unpaid research and development department? It's called the EU*. https://ar.al/2019/06/22/have-you-heard-about-silicon-valleys-unpaid-research-and-development-department-its-called-the-eu/.

Balkan, A (2019b). *Small Technology*. https://ar.al/2019/03/04/small-technology/.

Banning, M. E. (2015). Shared entanglements – Web 2.0, info-liberalism & digital sharing. *Information Communication and Society, 19*(4):1–15.

Baricco, A. (2018). *The Game*. Einaudi.

Batty, M., Manley, E., Milton, R., & Reades, J. (2013). Smart London. In S. Bell, & J. Paskins (Eds.), *Imagining the future city*. Ubiquity Press.

Bendiek, A., Godehardt, N., & Schulze, D. (2019). The age of digital geopolitics. *International Politics and Society*. https://www.ips-journal.eu/in-focus/chinas-new-power/article/show/the-age-of-digital-geopolitics-3593/.

Berners-Lee, T. (2017, March 12) I invented the web. Here are three things we need to change to save it. *The Guardian*. https://www.theguardian.com/technology/2017/mar/11/tim-berners-lee-web-inventor-save-internet.

Bria, F. (2015). Growing a digital social innovation ecosystem for Europe. *Digital social innovation final report*. https://www.nesta.org.uk/sites/default/files/dsireport.pdf.

Bromley, H. (1998). Introduction: Data-driven democracy? Social assessment of educational computing. In *IndieWeb*. https://indieweb.org.

Cadwalladr, C. (2013). Stewart Brand's Whole Earth Catalog, the book that changed the world. *The Guardian*. https://www.theguardian.com/books/2013/may/05/stewart-brand-whole-earth-cataloge.

Cambridge Dictionary. (2020). *Clickactivism*. https://dictionary.cambridge.org/it/dizionario/inglese/clicktivism.

Caulier-Grice J., Davis A., Patrick R., & Norman, W. (2012). *"Defining Social Innovation", TEPSIE*. European Commission Brussels.

Charalabidis, Y., & Koussouris, S. (2014). *Empowering open and collaborative governance: Technologies and methods for online citizen engagement in public policy making*. Springer-Verlag.

Cities for Digital Rights Declaration. (2020). *About*. https://citiesfordigitalrights.org/about.

Coleman, E. G. (2013). *Coding freedom: The ethics and aesthetics of hacking*. Princeton University Press.

Coleman, E. G. (2020). *Hack_Curio*. https://hackcur.io/.

Coleman, G. (2014). *Hacker, Hoaxer, Whistleblower, Spy: The story of anonymous*. Verso.

Commons Network. (2020). https://www.commonsnetwork.org/programme/commons-for-europe-2019/.

Communia. (2019). *Internet is for the people*. https://reform.communia-association.org/.

Coenegrachts, K. F. (2019). *Saving the internet for the many not the few*. https://citiesofpeople.com/en/2019/01/16/saving-the-internet/.

Coenegrachts, K. F. (2020). *Cities of people. Recipes for cities of the future*. https://citiesofpeople.com/en/citiesofpeople-2/citiesofpeople.

Cornwall, A. (2008). Unpacking 'participation': Models, meanings and practices. *Community Development. Journal, 43*(3), 269–283.

Cottica, A. (2010). *Wikicrazia L'azione di governo al tempo della rete. Capirla, progettarla, viverla da protagonista*. Navarra Editore.

Datta, A., & Odendaal, N. (2019). Smart cities and the banality of power. *Environment and Planning D: Society and Space, 37*(3), 387–392.

Davies, A., & Simon, J. (2012). *The value and role of citizen engagement in social innovation*. A deliverable of the project: "The theoretical, empirical and policy foundations for building social innovation in Europe, in: (TEPSIE), European Commission –7th Framework Programme, Brussels: European Commission.

de Rosnay, M. D. (2006). Alternative policies for alternative Internets. *Journal of Peer Production, 9*, 1–10.

de Souza e Silvas, A. (2006). From cyber to hybrid: Mobile technologies as interfaces of hybrid spaces. *Space and Culture, 9*(3), 261–278.

de Waal, M. (2014). *The city as interface: How new media are changing the city*. NAi010 Publishers.

DECODE. (2020). *What is decode?* https://decodeproject.eu/what-decode.

DW. (2019, November 25). *Web inventor Tim Berners-Lee unveils plan to save the internet*. https://www.dw.com/en/web-inventor-tim-berners-lee-unveils-plan-to-save-the-internet/a-51395985.

EC. (2009). *Regulatory framework for electronic communications in the European Union. Situation in December 2019*. Publications Office of the European Union. https://ec.europa.eu/digital-single-market/en/news/regulatory-framework-electronic-communications-european-union.

EC. (2013). *Future internet public–private partnership*. https://ec.europa.eu/digital-single-market/en/news/future-internet-%E2%80%93-public-private-partnership-fi-ppp-0.

Fritze, M. P., Gänser-Stickler, G. M., Türk, S., & Zhao, Y. (2019, October 28). F*ck off Google": Protest against Google Campus Berlin. *The CASE Journal*. https://www.emerald.com/insight/content/doi/10.1108/TCJ-03-2019-0013/full/html?skipTracking=true.

Graham, M. (2014). Inequitable distributions in internet geographies: The global south is gaining access, but lags in local content. *Innovations: Tecnology, Governance, Globalization, 9*(3–4): 3–19.

Graham, S., & Marvin, S. (2001). *Splintering urbanism: Networked infrastructures, technological mobilities and the urban condition*. Routledge.

Guijt, I., & Shah, M. K. (1998). *The myth of community: Gender issues in participatory development*. Intermediate Technology Publications.

Hague, C., & Williamson, B. (2009). *Digital participation, digital literacy, and school subjects: A review of the policies, literature and evidence*. Futurelab.

Harris, J. (2018, February 1). The punk-rock Internet. *The Guardian*. https://www.theguardian.com/technology/2018/feb/01/punk-rock-Internetdiy-rebels-working-replace-tech-giants-snooperscharter.

Hunsinger, J., Allen, M. M., & Klastrup, L. (Eds.). (2019). *Second international handbook of Internet research*. Springer.

Jenkins H., & Thorburn, D. (2003). The digital revolution, the informed citizen, and the culture of democracy. In H. Jenkins, & D. Thorburn (Eds.), *Democracy and new media*. MIT Press.

Kaika, M., & Swyngedouw, E. (2000). Fetishizing the modern city: The phantasmagoria of urban technological networks. *International Journal of Urban and Regional Research, 24*(1), 120–138.

Kalbag, L. (2015, March 26). *The illusion of free. A list apart*. https://alistapart.com/column/the-illusion-of-free/.

Kalbag, L. (2017, November 09*). Planning for accessibility. A list apart*. https://alistapart.com/article/planning-for-accessibility/.

Keanbirch. (2019, March 26). The in/humanity of technoscience: Environments, imaginaries, and inequalities. *Political Economy of Technoscience*. https://politicaleconomyoftechnoscience.wordpress.com/author/keanbirch/.

Kitchin, R. (2014). The real-time city? *Big Data and Smart Urbanism. Geojournal, 79*(1), 1–14.

Komninos, N., Pallot, M., & Schaffers, H. (2012). Special issue on smart cities and the future internet in Europe. *Journal of Knowledge Economics, 4*(2): 119–134.

Lanier, J. (2006). Digital Maoism: The hazards of the new online collectivism. *Edge Magazine*. https://www.edge.org/conversation/jaron_lanier-digital-maoism-the-hazards-of-the-new-online-collectivism.

Ozman M., & Gossart C., (2018, September). Digital social Iinnovation: Exploring an emerging field. *ISIRC 2018 Conference* (pp. 5–7). Heidelberg. Available at SSRN: https://ssrn.com/abstract=3434363.

Ledwith, M., & Springett, J. (2010). *Participatory practice: Community-based action for transformative change*. Policy Press.

Loukis, E., Charalabidis, Y., & Androutsopoulou, A. (2017). Promoting open innovation in the public sector through social media monitoring. *Government Information Quarterly, 34*(1), 99–109.

Martınez, M., & Rosende, S. (2011). Citizen participation in local agendas 21: Critical questions of urban governance. *Scripta Nova, 15*(3), 1–11.

Maxwell, W. (2014). Technology neutrality in Internet, telecoms and data protection regulation. *Global media and communication watch 2014*. https://www.hlmediacomms.com/2014/11/17/technology-neutrality-in-internet-telecoms-and-data-protection-regulation/.

Mazzucato M. (2019, October 2). *Preventing digital feudalism*. Project Syndicate. https://www.project-syndicate.org/commentary/platform-economy-digital-feudalism-by-mariana-mazzucato-2019-10.

McLaren, D., & Agyeman, J. (2015). *Sharing cities: A case for truly smart and sustainable cties*. MIT Press.

McLean, J. (2020). *Changing digital geographies: Technologies, environments and people*. Palgrave.

Mohan, G. (2001). Beyond participation: Strategies for deeper empowerment. In B. Cooke, & U. Kothari (Eds.), *Participation, The new tyranny*. Zed Books.

Naughton, J. (2019, January 20). The goal is to automate us': Welcome to the age of surveillance capitalism. *The Guardian*. https://www.theguardian.com/technology/2019/jan/20/shoshana-zuboff-age-of-surveillance-capitalism-google-facebook.

Nielsen, J. (2006, October 9). The 90-9-1 rule for participation inequality. Social media and online communities. *Nielsen Norman Group*. https://www.nngroup.com/articles/participation-inequality/.

OECD. (2020). *Internet policy and governance*. https://www.oecd.org/sti/ieconomy/internet-policy-and-governance.htm.

Parayil, G. (2005). *Information capitalism*. Springer SCOOP.IT. https://www.scoop.it/.

Platteau, J. (2008). *Pitfalls of participatory development*. Chapter prepared for the United Nations.

Raymond, M. (2012). The internet as a global commons? Governing the internet: Chaos, control or consensus? *Centre for international governance innovation*. https://www.cigionline.org/publications/internet-global-commons.

Sample, I. (2019, November 24). Tim Berners-Lee unveils global plan to save the web. *The Guardian*. https://www.theguardian.com/technology/2019/nov/24/tim-berners-lee-unveils-global-plan-to-save-the-internet.

Sennett, R. (2012, December 4) No one likes a city that's too smart. *The Guardian*.

Shaw, J., & Graham, M. (2017). An informational right to the city. In J. Shaw & M. Graham (Eds.), *Our digital right to the city*. Meatspace Press.

Sheppard, E. (2004). The spatiality of the limits to capital. *Antipode, 36*(3), 470–479.

Small Tech Foundation. (2020). *About*. https://small-tech.org/about/.

Söderström, O., Paasche, T., & Klauser, F. (2014). Smart cities as corporate storytelling. *City, 18*(3), 307–320.

Soulard F. (2018, October 31). The internet as a common good: Framework and perspectives for a citizen internet. *Science & Technology*. https://wsimag.com/science-and-technology/44147-the-internet-as-a-common-good.

Stokes, M. (2020). *What system factors help DSI to grow and thrive?* DSI4EU, Nesta. https://digitalsocial.eu/blog/145/a-first-run-of-the-dsi-index.

Streshinsky, M. (2018, October). *Stewart brand and the tools that will make the whole earth better*. Wired. https://www.wired.com/story/wired25-stewart-brand-whole-earth/.

Traldi, L., & Bria, F. (2018). *Design@large*. https://www.designatlarge.it/barcelona-smart-city-francesca-brmia-participatory-deocracy/?lang=en2018.

Turner, F. (2006). *From counterculture to cyberculture*. University of Chicago.

Vanolo, A. (2013). Smartmentality: The smart city as disciplinary strategy. *Urban Studies, 51*(5), 883–898.

Volpicelli, G. M. (2019, June 20). The obscure London startup that made Facebook's Libra possible. *Wired*. https://www.wired.co.uk/article/facebook-libra-startup-privacy-analysis.

Westera, W. (2012). *The digital turn. How the internet transforms our existence*. http://www.thedigitalturn.co.uk/TheDigitalTurn.pdf.

Wiener, A. (2018, November 16) The complicated legacy of Stewart Brand's "whole earth catalog". *The New Yorker*. https://www.newyorker.com/news/letter-from-silicon-valley/the-complicated-legacy-of-stewart-brands-whole-earth-catalog.

Zhao, Y., Pugh, K., Sheldon, S., & Byers, J. L. (2002). Conditions for classroom technology innovations. *Teachers College Record, 104*(3), 482–515.

Zuboff, S. (2019). *The age of surveillance capitalism: The fight for a human future at the new frontier of power: Profile Books*. PublicAffairs.

Links

Centre for International Governance Innovation. https://www.cigionline.org/.

Chainspace. https://chainspace.io/.

DECODE. https://decodeproject.eu/.

Diaspora. https://diaspora.social/.

IndieWeb. https://indieweb.org/.

MaidSafe. https://maidsafe.net/.

Mastodon. https://www.maastodon.net/.
Safe Access for Everyone. https://safenetwork.org/.
Small Technology Foundation. https://small-tech.org/.
Social Linked Data –Solid. https://solid.mit.edu/.
Steemit. https://steemit.com/trending/beos.
WikiLeaks. https://wikileaks.org/.

Index

C. Certomà, *Digital Social Innovation*,
https://doi.org/10.1007/978-3-030-80451-0

The manufacturer's authorised representative in the EU is Springer Nature Customer Service Centre GmbH, Europaplatz 3, 69115 Heidelberg, Germany. If you have any concerns regarding our products, please contact ProductSafety@springernature.com

Printed and bound by CPI Group (UK) Ltd, Croydon, CR0 4YY
22/09/2026
02221066-0020